Innovative Practices in Teaching Information Sciences and Technology

John M. Carroll
Editor

Innovative Practices in Teaching Information Sciences and Technology

Further Experience Reports and Reflections

Springer

Editor
John M. Carroll
College of Information Sciences and
Technology
Pennsylvania State University,
University Park
Pennsylvania, PA, USA

ISBN 978-3-031-61289-3 ISBN 978-3-031-61290-9 (eBook)
https://doi.org/10.1007/978-3-031-61290-9

This Springer imprint is published by the registered company Springer Nature Switzerland AG
The registered company address is: Gewerbestrasse 11, 6330 Cham, Switzerland

Contents

Teaching as Innovation

John M. Carroll

Abstract Innovative teaching can be achieved in many ways—leveraging what students already do, know about, and think about; relying on student engagement, self-regulated activity, and collaborative activity; and embedding teaching discourses in the values, practices, and institutions of the domain. Teaching innovation is especially critical and appropriate in computer and information sciences: *critical* because of the need to enroll and graduate more students and stronger students in these areas and *appropriate* because these areas broadly depend upon and consist of continuous innovation. This book describes and reflects on a diverse collection of innovative teaching practices across the faculty of the College of Information Sciences and Technology at Pennsylvania State University.

Keywords Teaching practice · Innovative teaching · Information sciences and technology

Teaching may be the original and most important innovation in human history. Sharing elements of knowledge and skill with others, guiding and inspiring them to assimilate and make sense, to practice and apply, to appropriate and re-appropriate, and to criticize, reinterpret, and extend, is the very essence of human progress and essential to human existence. And to the converse, whatever insights and techniques are not shared are not discussed and debated, are not applied and explained, and ipso facto never have impact; they never make a difference.

The key activities that comprise teaching are all about innovation. Young children are easily motivated to learn in order to make sense of and participate in everyday community activity (Dewey, 1938). It isn't *so* different for more developed humans. What is different is that they have already learned enough to be intentional about their motivation. They have more articulate ideas about what makes sense and how

J. M. Carroll (✉)
College of Information Sciences and Technology, Pennsylvania State University, University Park, Pennsylvania, PA, USA
e-mail: jmc56@psu.edu

J. M. Carroll (ed.), *Innovative Practices in Teaching Information Sciences and Technology*, https://doi.org/10.1007/978-3-031-61290-9_1

they wish to participate. This can make adults more discerning, more capable, and more focused learners, but it can also make them skeptical about engaging with teaching and learning merely because another adult encourages them to do so. Exciting students about knowledge and skill requires continual innovation in teaching. For only when learning makes sense, when it addresses issues and questions people understand and value, and when it matters to their communities and societies can teaching soar.

Innovative Teaching

Motivation to learn is the essential prerequisite; it opens students to teaching. But the subsequent interactions of teaching must continue to strengthen and enhance motivation. No one now thinks of teaching as merely pouring knowledge into a student's (presumably empty) head with the Nurnberg Funnel (a scary German legend about teaching in the Middle Ages, and the title of my first book on learning, Carroll, 1990). But we do sometimes imagine that the key vehicle of teaching is a well-structured slide presentation. It is of course an important idea to organize content deliberately, but structure per se is not a sufficient goal. Imparting knowledge and skill to students must raise and address significant questions so as to engage, provoke, and develop their minds and their imaginations. That's teaching. An organized presentation of content can play a role in this but is never sufficient.

What else is required for innovative teaching? I am confident that there is not a single recipe for engaging students and helping them learn. That is one reason why there should be more books (like this one) in which teachers describe and reflect on how they achieved teaching innovations. Four of the most general and powerful ideas about innovative teaching are enumerated immediately below. For the most part, these are ideas I first encountered empirically as a young IBM researcher carrying out studies of office workers trying to learn to use then-novel word processors (Mack et al., 1983). I later developed the concept of *minimalist instruction* based on that work (Carroll, 1990, 1998). Subsequently, and for the past 30 years, I have been a university professor. Initially, I tried teaching with well-structured slide presentations, but eventually I adopted my own ideas (Carroll et al., 2015, 2016; Carroll & Rosson, 2005; Du et al., 2012; Rosson & Carroll, 2013).

1. *Learning should be* meaningful *to students; it should leverage students' prior knowledge and experience and issues and questions that students already understand.*

Learning is typically making new meanings by reshaping and extending what is already familiar. New information, interactions, and experiences are incorporated into what is already known; at the same time, existing knowledge and skill changes and develops to accommodate and better exploit new insights, new ways of seeing, and new ways of doing (Bruner, 1966; Piaget, 1985; Vygotsky, 1978).

The office workers we studied long ago at IBM Research used their knowledge of office objects and work process and their skills of typing with typewriters (remember, this was the early 1980s) to scaffold their learning about word processors and the office work processes entailed by that technology. No one had to suggest this strategy to them, and in some cases the strategy caused them problems. For example, the Backspace command in a word processor does not just move the typing point backward; it *deletes the user's data* as it goes. This caused some alarm among our participants, and one might rightfully question the design of the Backspace command (which in many keyboards is now renamed Delete), but the behavior of our participants is a normal and robust example of how people leverage what they already know to make meaning in what they are learning.

2. *Student roles in learning activity should include* self-directed reasoning and improvisation.

Learners should be creative and active participants, not just witnesses in their own learning (Dewey, 1938; Vygotsky, 1978). Thus, in leveraging what students already know and do in order to make learning more meaningful to them, teachers should always consider empowering students to *make meaning*, rather than just hearing or reading about meaning. A routine example is instead of reviewing the content of a reading assignment for the class, a teacher might pose questions about it or invite students to compare and contrast readings with respect to given themes and issues. The latter are *active learning* strategies (Bruner, 1966; Dewey, 1938).

3. *Students should be involved in* realistic activity*; applying knowledge in authentic contexts is the best way to encounter and acquire knowledge. Realistic activity is also* collaborative*; students should work and learn together, as people do in the real world.*

The context of learning is part of what is learned. When learning occurs in austere academic contexts, such as lectures or rote exercises, what is learned will be better remembered in lecture halls, but less accessible to learners in project-based work activity or in debates about societal values (Choi et al., 2023). But this is the opposite of what people need from teaching and learning. Since students typically know little about real contexts, such as work activity and debating values, it makes sense to incorporate realistic activity contexts with the events of teaching and learning whenever possible. This would allow students to learn how knowledge and skill can be useful as they first learn what it is.

Action-oriented teaching and learning is most effective when it is authentically embedded in the values, practices, and institutions of the subject domain (Lave & Wenger, 1991). For example, as an undergraduate student in mathematics (in the late 1960s), I proved many theorems, but I did this mechanically to work through problem sets, and I was careful never to talk to the other students. This is not how mathematicians work. They prove theorems strategically to build tools that enable further results and questions, and they talk to one another about what they are doing. In this sense, I feel that while much of my mathematics education was active learning, I did not engage mathematics authentically. At the end of my degree, I still had only the roughest idea of what it would mean to *do* mathematics. This is an inadequate educational outcome.

4. *Teaching should legitimize and* make constructive use of error and misconception*; errors are inevitable if students think on their own and are intrinsically motivating for students to diagnose and debug and to recover from.*

Much learning in the real world is driven by *breakdown*, Heidegger's term for things not working as expected (Koschmann et al., 1998). When one's knowledge and skill produce errors instead of results, one turns to learning and innovation. Breakdown can be uncomfortable, even paralyzing, but it can also be riveting and productive when it is experienced as presenting new challenges.

Students confront breakdown regularly because they are learners and necessarily exploring the boundaries of concepts and strategies. Teachers can help students experience their errors and misconceptions as actionable and useful, not as failures. They can convey to students that understanding and recovering from errors is an important path to learning (Jackson et al., 2022). This can help students to be more successful students and, throughout their working lives, to be better at managing and benefitting from breakdowns. Empowering and challenging students to think critically about the world as it is and to accept responsibility for the future is innovation.

Information Sciences and Technology

This book is about teaching innovation in the College of Information Sciences and Technology at the Pennsylvania State University, hereafter referred to IST. IST is an interdisciplinary college including artificial intelligence, data science and applications, human-centered informatics, information infrastructures and practices, and social and organizational informatics. The college was founded in 1999 with five faculty; it has grown to more than 80 full-time faculty and more than 2000 undergraduate students, nearly 300 graduate students, and more than 1000 online students.

The academic terrain of IST is an example of an *iSchool*; an international group of more than 120 universities now have iSchools, addressing "all aspects of research and teaching about information" (https://www.ischools.org). The iSchool movement coalesced in the early 2000s, a period when undergraduate enrollments in computer science were plummeting (Computing Research Association, 2013). In that context, iSchools sought to establish and address a broader vision of information as a lynchpin in human activity and in the future of universities. For iSchools, human concerns and possibilities are at the center of research and teaching about information technology, affording more integrated innovation of technology, social responsibility, and human experience. No surprise, faculty in iSchools typically represent a wide range of disciplinary origins and perspectives, computer and information science, psychology and sociology, philosophy, economics, and political science, as well as business, law and engineering, and people who themselves were educated in interdisciplinary programs.

Part of the vision of iSchools is that broader academic programs in information technology and innovative teaching by interdisciplinary faculties will attract and retain a larger and more diverse student population. The need and the opportunity for this are significant: five of the seven entry-level job categories with the highest projected need, highest growth rate, and highest salary for the United States are from general area of computer and information technology (Bureau of Labor Statistics, 2022). And this was *before* the 2023 boom in applications of generative artificial intelligence. All this suggests that the iSchool area will continue to develop.

Teaching innovation was a founding commitment of the IST faculty. In the early years of IST, many faculty gathered for a lunchtime meeting about innovative teaching. This was never a top-down directive, but it was encouraged by the enthusiasm of Larry Spence, our Director of Teaching and Learning in the college, by faculty peer interactions, and by insights and results in classes. When I interviewed for a job in IST in 2003, I experienced it as having a culture of teaching innovation, a shared value. It is a special possibility of an academic faculty to create such a culture, and I am grateful and proud to have participated in it. I joined IST partly because of this culture. I still recall my surprise that several faculty members I met on my job interview (including Larry Spence) were familiar with my prior work on learning (which was not the primary research area I was presenting at my interview) and wanted to talk about innovative teaching.

This book focuses on the innovative teaching practices of a single iSchool faculty. Thus, this collection of experience reports is more a faculty case study than the sort of convenience survey one generally sees in an edited collection. I couch the distinction this way because, in discussing this project with several colleagues, it seemed one reaction was to question the validity of concentrating on a single faculty. I think this is a serious point, but it needs to be considered in the context of standard practices, which often hinge on over-sampling members of the editor's professional network. Thus, even though the authors of the typical edited collection appear to be a diverse group, in some senses they are handpicked (by the editor) to be likeminded.

Moreover, I think that focusing on a single faculty case study allows recognition and acknowledgment of the thesis that teaching innovation is an issue of faculty culture, not merely a matter of individual instructor choice and initiative. I think this is also a serious point regarding what teaching innovation is, how it emerges, and how it can be sustained. The typical approach of reporting and examining teaching innovations course-by-course or instructor-by-instructor and sampling widely across universities and countries does not address this question.

This Book

This book is the second of two. In 2014, my IST colleagues and I (including faculty, staff, and students) published the 18 chapters of *Innovative Practices in*

Teaching Information Sciences and Technology: Experience Reports and Reflections (Carroll, 2014). Like that previous book, the chapters of this book are experience reports and reflections on teaching practices of IST faculty, staff, and students. Each essay describes specific pedagogical challenges and goals that a particular instructor addressed through her or his personal educational vision and values. We have tried to keep these reports succinct and concrete, in some cases pointing to other publications and web resources. The objective was to be inspiring, but more importantly to be grounded enough that other faculty could do something, emulate, adapt, and appropriate a technique or approach in their own courses.

The first five chapters address perennial challenges for teaching in IST. In the chapter "The False Promises of Application-Driven Learning: Mathematical Thinking in Today's Rapidly Evolving Technology Landscape," Sarah Rajtmajer observes that although IST undergraduates have selected a degree program with foundations in mathematics, they can still experience math anxiety. Based on her experience teaching a required math course for IST undergraduates, Rajtmajer shares strategies and perspectives. She argues that framing math as justified only by its applications can deprive students of the opportunity to overcome their anxiety and cultivate self-efficacy.

Shomir Wilson seeks to enable broader use of class discussions in the chapter "Professor in a Strange Land: An Expat Metaphor for Classroom Discussions". Wilson suggests that faculty embrace the generational gap between them and their students; the instructor is an "expat" immersed in a class consisting of people who understand, use, and experience technology differently. The metaphor converts what might be just a communication challenge into a collaborative resource, legitimizing the expertise students bring to the class and motivating their opportunity and responsibility to contribute perspectives.

Learning a programming language is learning a huge number of details that can obscure more general structures and concepts of the language. In the chapter "When Less Is More: A Java Introductory Course," Fred Fonseca describes his efforts to simplify initial presentation of language details more rigorously in order to help students keep general structures and concepts in mind as they learn.

Chris Gamrat addresses the related issue of introducing technological concepts to new and prospective IST students. In the chapter "Teaching Information Technology Without the Technology: An Unplugged Approach to Introducing IT Concepts to First-Year Students," he shows how paper presentations of technologies like social network analysis, programming, and cryptography can be effective and easier to manage.

In the chapter "Scenario-Based Design with ChatGPT: A First-Year Seminar Experience," John Carroll describes enrichment for a first-year seminar, a type of course primarily concerned with orienting new students to resources and opportunities of the campus and student life. This chapter discusses how IST method and tool content suggested new activities for first-year topics such as academic integrity and allowed students to carry out more meaningful course projects.

The next four chapters address emerging challenges for teaching in IST. Machine learning applications have become pervasive through the past decade, and many college students without strong backgrounds in computer science and mathematics are interested in learning how to use machine learning. In the chapter "Hitting the Sweet Spot: Designing an Accessible Introductory Machine Learning Course for Informatics Students," Amulya Yadav and Hangzhi Guo take up this challenge, developing a use-oriented approach to an introductory course in machine learning.

Lizhen Zhu and James Wang describe the challenge that most data science education occurs in graduate programs, leveraging and integrating existing expertise of students. In the chapter "Unleashing the Problem-Solving Potential of Next-Generation Data Scientists," they discuss developing a project-focused, writing-intensive course for upper-level undergraduates in which students experience all phases of a data science project.

In the chapter "Providing Iterative Cybersecurity Hands-on Learning Experience: Reflections on Teaching Cyber-Defense to Second Year IT Students," Peng Lui critically reconsiders hands-on cybersecurity labs and describes course development that iteratively improved learning dependencies throughout a sequence of labs. This approach supported sequential mastery of advanced cybersecurity skills.

Another emerging challenge for teaching in IST is explainable Artificial Intelligence. In the chapter "Learning About Explainable AI with Very Little Programming For 4th Year Undergrads (or Younger)," Jonathan Dodge, Antoinette Diawuo, Enyan Dai, and Rupak Das describe the development of an undergraduate course designed to be "accessible to fledgling data scientists." For example, they minimized course dependencies on programming, mathematics, and human-computer interaction.

The next four chapters address embedding the classroom in the world. Dan Welch, in the chapter "Traversing the Software Development Life Cycle: Reflections on Teaching Object-Oriented Programming to Third Year IT Students," leverages the software development lifecycle as a template for teaching and learning software concepts and skills. Students enact the process model to learn about it and to learn about the structures, techniques, and tools that comprise it. For example, version control is employed throughout the course.

Guoray Cai's chapter "Blurring the Boundaries of Teaching Modes to Improve the Experience of Professional Learners" notes that a by-product of the COVID pandemic was instructional development focused on various modes (face-to-face, synchronous remote, asynchronous, and hybrid). Cai describes how teaching modes can be more seamlessly combined for Penn State World Campus students (whose default mode is asynchronous) to produce better outcomes.

In the chapter "Semester Projects on Human-Computer Interaction as Service and Outreach, undergraduate and graduate," Frank Ritter and Sarah Stager argue that course projects can be more engaging and useful to students if they are more real, for example, having specific social and economic impact, in addition to providing students opportunities to integrate and apply course concepts and techniques.

Mahir Akgun examines the role of knowledge creation in learning in the chapter "Knowledge Integration and Knowledge Building: Engaging with Different Perspectives to Create a Place for Knowledge Building in a Learning Environment." Based on experience with a visual analytics course, he argues that incorporating knowledge building into learning activity supports a more dynamic, collaborative, and stimulating environment.

The final four chapters address management and development issues for courses and curricula. In the chapter "Single Point Rubrics: An Opportunity in Graduate Education," Edward J. Glantz and Ralph J. Osmolinski critically review grading rubrics, explicit models for allocating credit that provide transparent subgoal guidance to students in carrying out assignments and help explain grading outcomes. They focus especially on using rubrics in graduate courses.

David Hozza, in the chapter "Entering the Cybersecurity Workforce: Certification vs. College Degree," describes the emergence of the cybersecurity profession and current challenges to train sufficient numbers of qualified cybersecurity professionals to meet the needs of industry and government.

In the chapter "Teaching as a Practice of Cultivating Relationships," Priya Kumar, Yao Lyu, and Ananya Reddy, a faculty member, a graduate teaching assistant, and an undergraduate student, respectively, describe teaching as relationship building. They illustrate four principles of relationship-rich education through their personal teaching and learning experiences.

Rosalie J. Ocker and Lisa Lenze, in the chapter "Achieving Consistent and Relevant Learning Outcomes on a Common Course," discuss the redesign of an undergraduate core course in database design and development. The course is required for most programs in the IST college and is taught by a wide range of faculty each year. Ocker and Lenze's project sought to create a common course design that would satisfy diverse students and faculty across programs in IST.

I joined IST in 2003. I was a bit past mid-career. I believe I had always tried to be innovative in my teaching, but I appreciated and was inspired and encouraged by the collegial validation and support for this in IST. I think it helped me to become more innovative and to keep trying to be innovative more than I might have done on my own.

Though I was motivated to organize this book by a sense that many of my colleagues were innovative teachers, I have ended up impressed at just how innovative they are. Strategies like leveraging what students do, know about, and think about, relying more on discussion and activity than on lecture and description, employing collaborative learning, designing course activities that are authentic to real-world domains, and using signature projects and practices are not exceptional or radical practices in our college; they are teaching norms. This is quite exciting. It was not what I experienced myself as a student, nor is it what I saw when I first became a faculty member.

This book is not a claim that my colleagues and I are more innovative than other faculty. Rather, it is an acknowledgment that the culture of teaching and learning in a faculty is more than the practice of any faculty member. This collection of reflective essays on teaching practices is a vehicle to begin exploring that proposition and, more specifically, to reflect on and extend innovative teaching in IST, to continue to learn from one another, to help one another achieve more. This book is a vehicle to share a collection of innovative practices and experiences and IST's continuing culture of teaching innovation with other colleagues. It is a celebration of innovative teaching.

Acknowledgments I am grateful to my innovative teaching colleagues for encouraging the concept for this book (and its predecessor) and for working to develop and share that concept. More personally, I hope this honors the memory of my former colleague Professor Larry Spence who did more than anyone to establish the culture of innovative teaching and learning in our college.

References

Bruner, J. S. (1966). *Toward a theory of instruction*. Belknap Press/Harvard University Press.

Bureau of Labor Statistics. (2022). *Occupational outlook handbook*. At http://www.bls.gov/ooh/. Accessed 11 Feb 2024.

Carroll, J. M. (1990). *The Nurnberg Funnel: Designing minimalist instruction for practical computer skill*. MIT Press.

Carroll, J. M. (Ed.). (1998). *Minimalism beyond "The Nurnberg Funnel"*. M.I.T. Press.

Carroll, J. M. (Ed.). (2014). *Innovative practices in teaching information sciences and technology: Experience reports and reflections*. Springer.

Carroll, J. M., & Rosson, M. B. (2005). A case library for teaching usability engineering: Design rationale, development, and classroom experience. *ACM Journal of Educational Resources in Computing, 5*(1), Article 3, 1–22.

Carroll, J. M., Jiang, H., & Borge, M. (2015). Distributed collaborative homework activities in a problem-based usability engineering course. *Education and Information Technologies, 20*(3), 589–617.

Carroll, J. M., Wu, Y., Shih, P. C., & Zheng, S. (2016). Re-appropriating a question/answer system to support dialectical constructivist learning activity. *Educational Technology Research and Development, 64*(1), 137–156. http://link.springer.com/article/10.1007%2Fs11423-015-9405-6

Choi, Y., Kim, K., Moon, S., Jung, G., Cha, J., & Yim, H. (2023). Context dependent memory in the wilds. *Proceedings of the Annual Meeting of the Cognitive Science Society, 45*(45). Retrieved from https://escholarship.org/uc/item/8xh0f8gx

Computing Research Association. (2013). *Annual Taulbee Survey*. At http://cra.org/resources/taulbee/. Accessed 2 Mar 2024.

Dewey, J. (1938). *Experience & education*. Kappa Delta Pi.

Du, H., Rosson, M. B. & Carroll, J. M. (2012). Augmenting classroom participation through public backchannels. *ACM Group 2012: International conference on supporting group work*, (October 28–31, Sanibel Island, FL). ACM, pp. 155–164.

Jackson, A., Godwin, A., Bartholomew, S., & Mentzer, N. (2022). Learning from failure: A systematized review. *International Journal of Technology and Design Education, 32*(3), 1853–1873.

Koschmann, T., Kuutti, K., & Hickman, L. (1998). The concept of breakdown in Heidegger, Leont'ev, and Dewey and its implications for education. *Mind, Culture, and Activity, 5*(1), 25–41.

Lave, J., & Wenger, E. (1991). *Situated learning: Legitimate peripheral participation*. Cambridge University Press.

Mack, R. L., Lewis, C. H., & Carroll, J. M. (1983). Learning to use office systems: Problems and prospects. *ACM Transactions on Office Information Systems, 1*, 254–271.

Piaget, J. (1985). *The equilibration of cognitive structures: The central problem of intellectual development*. University of Chicago Press.

Rosson, M. B., & Carroll, J. M. (2013). Developing an online community for women in computer and information science: A design rationale analysis. *AIS Transactions on Human-Computer Interaction, 5*(1), 6–27.

Vygotsky, L. S. (1978). *Mind in society: The development of higher mental processes*. Harvard University Press.

Part I
Meeting Students Where They Are: Perennial Challenges

The False Promises of Application-Driven Learning: Mathematical Thinking in Today's Rapidly Evolving Technology Landscape

Sarah Rajtmajer

Abstract Undergraduate curricula in informatics programs typically include several courses in mathematics. In my time teaching mathematics in IST, I have noticed continual and increasing calls from colleagues and administrators to prove to students the relevance of these courses through application-driven lesson plans and preferential treatment of topics with well-understood connection to today's technologies. This chapter offers a counterpoint to these calls. I suggest that the treatment of foundational mathematical concepts as a means to well-defined and highly circumscribed ends perpetuates math anxiety and undermines flexibility of mathematical reasoning critical to highly dynamic IST careers. The examples I provide center around a course on discrete mathematics. However, the message is general.

Keywords Discrete math · Math for information sciences and technology · Math anxiety

Introduction

By the time they reach mid-career, today's students of information sciences and technology (IST) will be engaging with technologies they did not learn about and could not have imagined as undergraduates—technologies, perhaps, powered by AI and quantum computing—recently in the realms of science fiction. A premise of this chapter is that the primary goal of an IST curriculum should be the provisioning of a rich set of skills and a growth mindset which students can draw upon throughout their careers as they learn, work, and innovate in a rapidly evolving technological landscape.

S. Rajtmajer (✉)
College of Information Sciences and Technology, The Pennsylvania State University, University Park, PA, USA
e-mail: smr48@psu.edu

J. M. Carroll (ed.), *Innovative Practices in Teaching Information Sciences and Technology*, https://doi.org/10.1007/978-3-031-61290-9_2

At the center of such a skillset, I argue, is mathematical reasoning. Some examples of mathematics in IST are ubiquitous: number theory for cryptography, set theory for password security, and relational algebra for databases. We pull out these examples when our students ask us "why they need to learn this." But if we are being honest, we know that public key cryptographic algorithms will be replaced soon (NIST Announces First Four Quantum-Resistant Cryptographic Algorithms, 2022) and passwords even sooner (Goodfellow et al., 2016). It turns out that deep learning has its foundations in algebra (Dweck, 2016). That is convenient for instructors tasked with teaching linear algebra and fielding students' questions about real-world utility. I wonder, what did they say before AI? And what will they say next?

My message brings together two challenges which on the surface appear unrelated: natural tensions between theoretical and applied mathematics and students' experiences of stress, anxiety, and lack of confidence in mathematics. My experience in the classroom is that these challenges are intertwined and mutually perpetuating.**I argue that overemphasis on the application of fundamental mathematical concepts to specific exemplars within today's technology landscape situates mathematics as a means to an end, encourages memorization rather than imagination, and undercuts students' confidence in their ability to engage with mathematical reasoning more broadly.** This will backfire, I fear, when students are faced with new ideas and new technologies down the road—an inevitability in the rapidly changing fields of IST.

This chapter will share suggestions for instructors in undergraduate-level mathematics classes for IST majors to mitigate math anxiety and promote a growth mindset (Dowker et al., 2016) through engagement with historical context, philosophy, puzzles, and games.

Math Anxiety

Decades of research in education and psychology have explored stress, anxiety, and lack of confidence in mathematics—so-called math anxiety (see Ashcraft, 2002 for an overview). School-age children have been the focus of much of this work, although researchers have suggested that large subsets of the general population at all ages are impacted.

The literature suggests that math anxiety is a phenomenon with both social and cognitive underpinnings (Tobias, 1993). Reflective of political and power structures underlying perceptions of intelligence, math anxiety is not felt equally by all. It impacts women and minorities more significantly (Nosek et al., 2002; Yu et al., 2024). Math anxiety is not only more pervasive and severe in women than in men but also believed to be more closely associated with performance in female students than in male students (Yu et al., 2024). Research thus far has failed to produce any evidence that sex differences in mathematics performance have biological basis, yet sociocultural context has perpetuated a notion that the male brain is more logical and

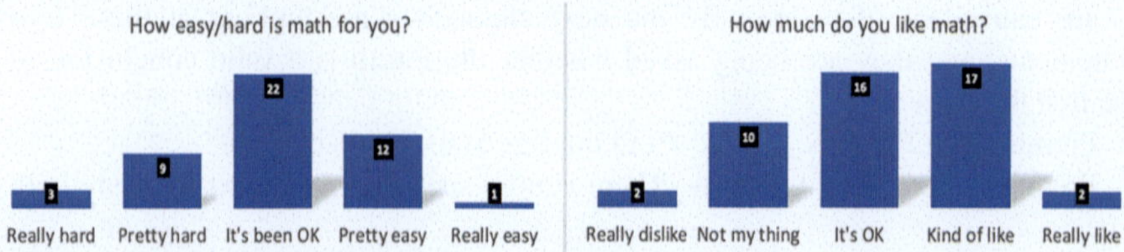

Fig. 1 Results of two-question in-class survey of IST 230 students

better at math (Mendick, 2005). This is a vicious cycle for women and girls—fearful that they will not understand and that they cannot do, they don't.

I have observed a surprising prevalence of math anxiety over several years as an instructor of a foundational (required) mathematics course for undergraduates studying IST—surprising, perhaps, because these students have self-selected into a technical major that leans heavily on mathematical foundations.

This semester, I made my first attempt to quantify this pattern among my own students. I administered a survey to my class of 47 students in IST 230—discrete mathematics on the first day of classes at Penn State. I asked them how hard they have found math in prior courses and how much they enjoy math. The results are in Fig. 1. Results were well-distributed but lower than ideal for students in IST. My goal, of course, is to have moved the needle on these answers by the last day of class and for students to feel that change well beyond this course and this program.

Figuring Out as Growth Mindset

As a woman with a PhD in mathematics and a faculty member in a College of Information Sciences and Technology, I have had the opportunity to teach both theoretical and applied math courses in traditional math departments and in the context of IST where I sit now. My experiences have reinforced the value of mathematical reasoning for both audiences, equally.

Studying topics in theoretical mathematics is an exercise in getting comfortable with discomfort—in figuring out. Consider the first topic we teach in IST 230, propositional logic. I show students a logical expression. It involves symbols like $\exists$ which is shorthand for "there exists" and $\forall$ which means "for all." Most students have never seen these symbols before and that fact alone is enough to cause some discomfort. But most will dutifully learn the translation of symbols to text and will recite those definitions readily on homework or exams.

Next, we string these symbols together into statements. For example, students are asked to understand whether the two statements:

$\neg\forall x\, P(x)$: "It is not true that $P(x)$ is true for all x."

and

$\exists x \neg P(x)$: "There exists an x such that $P(x)$ is not true."

are equivalent (they are). By the next class, they are looking at these four statements, and they are being asked whether the fourth is a valid conclusion of the first three premises:

Premise: $\forall x(P(x) \to S(x))$—$P(x)$ implies $S(x)$ for all x.

Premise: $\neg\exists x(Q(x) \wedge R(x))$—There is no x such that $Q(x)$ and $R(x)$ are both true.

Premise: $\forall x(\neg R(x) \to \neg S(x))$—Not $R(x)$ implies not $S(x)$ for all x.

Conclusion: $\forall x(P(x) \to \neg Q(x))$—$P(x)$ implies not $Q(x)$ for all x.

It looks intimidating, but they know that they can *figure it out*.

Our subsequent module on proofs is a great example of getting comfortable with discomfort and figuring out. Proof-writing is a notoriously hard topic where math anxieties peak. I tiptoe into these waters starting with proofs that, at first glance, seem easy enough. Here's one:

Theorem If $m \in \mathbb{Z}$ is even, then m^2 is even.

I teach students general techniques that they can apply. I teach them about *I teach students general techniques that they can apply. I teach them about*. I teach them that sometimes it helps to break down a problem into smaller parts, i.e., cases. However, these are all just tools that may or may not be helpful for their problem at hand. There is no step-by-step guide I can give to a student tasked with proving a theorem. They need to open their toolbox and pick one to try. And if that tool doesn't work as planned, they need to put it back and pick up another one.

It is precisely in this moment—when a student wrestles with a problem that scares them—that they sit in an opportunity for growth. If they can work through it, they now have one datapoint for their memory bank of a time when they looked at a daunting mathematical concept and figured it out. Next time they are in a similar situation, they are less likely to panic and more likely to start figuring.

Can't all this happen in applied math too? Yes, of course. However, a subtle (but I argue, important) distinction when centering the importance of application is the focus on functionality rather than understanding. Applied math serves to solve a known problem with a known answer in a known way. Our implicit message to the student is this:

Someone (really smart) already figured out that this is the set of steps you take to solve this IST-related problem. You should commit these steps to memory and then use them when you see that problem come up again on the test or, ideally, outside the classroom.

Unfortunately, chances are that when that problem shows up again outside the classroom, it will look different than it did in class. The more time passes, the more likely that is to be true. But even today, most real-world applications of concepts of discrete mathematics in IST are far more nuanced and complicated than what we can show in the classroom (more on this in a moment). The set of steps the student memorized will not apply. Rather, we hope that the way of thinking we taught them will apply. So, if we agree on that, let us agree that building that way of thinking should be the goal from the start.

The conceptual concerns I've presented about application-driven mindset may seem overly academic, but they are manifested in tangible ways when compounded with practical realities. All math courses in IST try to cover a lot in a limited time. At Penn State, IST 230 is the one-semester course tasked with teaching students (all of) discrete math. The topics "covered" on this list would amount to at least four semester-long courses in an undergraduate mathematics curriculum, e.g., logic, set theory, number theory, and graph theory. This, of course, means we need to pick and choose where we spend more time and where we give just a light touch. Our teaching philosophy and our expected outcomes guide these decisions. For example, the logic and proofs I mentioned earlier in this chapter would likely be first on the chopping block in an application-driven course. Those concepts underlie all of programming, but, with limited time and an application-driven focus, you might be better off skipping those.

In-Class Strategies

If I've convinced you thus far that a more theoretical flavor is a worthy goal, you might wonder what steps take in the classroom to engage in this way. Here, I share some ideas, grouped according to high-level strategies that each instructor might interpret based on their classroom dynamics and teaching style.

Normalize Being Stumped

Use history to inspire Students might be surprised to learn that the world's greatest mathematicians spend much of their careers tackling problems that may take them years to solve or which they may never solve at all. Share with them, e.g., the inspiring story of Fermat's last theorem (Ribenboim, 1979). In 1637, Pierre de Fermat scribbled a note in the margin of a page of his copy of Arithmetica, an Ancient Greek text written by mathematician Diophantus of Alexandria (c. 250 BE). The note states that the following equation is unsolvable with positive integer values of x, y, and z, for all n greater than 2:

$$x^n + y^n = z^n$$

Fermat writes (translated):

I have a truly marvelous demonstration of this proposition which this margin is too narrow to contain.

After 358 years and the career-long dedication of countless mathematicians, Fermat's last theorem was finally proved in 1994 by Andrew Wiles, a professor at Princeton (Wiles, 1995). There are many stories about landmark problems and the mathematicians who devoted themselves to them of similar intrigue and magnitude in the history of mathematics. These anecdotes can serve to humanize and even glorify not knowing, figuring out or even never figuring out, and creativity.

Students might be intrigued to learn the notion of undecidability, i.e., some problems simply cannot be solved using the axioms of mathematics. One might present, for example, the 2015 finding that an important and long-standing mathematical problem underlying fundamental questions in particle physics and quantum physics is provably unsolvable (Cubitt et al., 2015). That example also gives students very tangible picture of recent and ongoing advancement of the field of math (vs. math as a relic of the past).

Use today for context In this vein, students might be surprised to know that there are many important unsolved problems still today. And, in fact, new problems continue to emerge as the field moved forward. This is the essence of research. I tell my IST 230 students about The Millennium Prize Problems,[1] seven open problems in mathematics with $1 Million prizes now attached to their solutions. Several times, students have followed up with me in the next class or in office hours to discuss them. They independently read those problems and, at least for a moment, perhaps thought this way their opportunity to make their million dollars.

In the classroom, one might consider devoting 20–30 minutes to let students experiment with an open problem themselves. One does not need to let the students know from the outset that the problem is unsolved but pick something relatively straightforward and let them work through the process of figuring out. Coloring problems work well at this level, I have found, and could be placed within a module on graph theory. You might ask students, essentially, to prove or disprove the four-color theorem (Gonthier, 2008)—that is,

Theorem Given any separation of a plane into contiguous regions, the regions can be colored using at most four colors so that no two adjacent regions have the same color.

Give each student four colored pencils and ask them to figure out whether this statement is correct or incorrect. Let them work in groups and bang their heads against it for a while. When it is time, they might be thrilled to know that the four-color theorem was not proved until 1976, despite more than 100 years of effort and many false proofs and counterexamples. Even then, it was proved only with the support of software.

Incorporate Puzzles, Games, and Fun

The correlation between liking math and perceived aptitude was very high in the in-class survey I deployed this semester (referenced in Fig. 1). In fact, no student had more than a 1-point difference between their two responses. Hence, it might be true that helping students find math more enjoyable might also help to improve

[1] Here is the Millennium Prize Problem as webpage: https://www.claymath.org/millennium-problems/

confidence and reduce anxiety. This idea has also been suggested in prior work (Middleton & Spanias, 1999).

To this end, using puzzles and games in class is an intuitive idea. Discrete math, luckily, is ripe with these opportunities. There are also some wonderful websites that collect and categorize these games and wonderful volumes on recreation mathematics (Bogomolny, 2024). During the course module on Boolean logic, one might spend a class period having student teams create games with logic gates. With some guidance, teams can create tic-tac-toe or river crossing games, e.g., Ascher (1990). During the course module on probability, for example, one might reenact the Monty Hall problem (Gill, 2010) with real (small) prizes.

One source of a multitude of puzzles, stories, and fascinating ideas is Douglas Hofstadter's 1979 masterpiece Godel, Escher, Bach: An Eternal Golden Braid (Hofstadter, 1999). The work has been so influential, in fact, that there have been undergraduate courses at MIT, Stanford, Dartmouth, and elsewhere dedicated entirely to the text. Many of the concepts at the center of the book, e.g., logic and recursion, are central to discrete mathematics. But indeed, the book could no doubt fit in beautifully as a source of enrichment for any mathematics course in the IST curriculum. For student inclined toward art, music, literature, and philosophy, the book will be particularly meaningful. But it is also worth noting the focus of GEB on linking mind and machine. In that way, it is also germane to IST and promotes creative thinking about technology that I have heralded throughout the chapter.

Rethink Grading

Finally, I think it is important to address issues around grading, although, I myself have few answers yet in this regard. I have seen students' laser focus on grades undermine growth mindset and willingness to be wrong that I have argued we need to encourage. Worry about grades, of course, heightens anxiety. In retrospect, I wish that I had included a question on my class survey about grades in prior math classes. I assume that all three—past performance, perceived difficulty, and enjoyment—are closely intertwined for most students.

To some extent, students' focus on and anxieties around grading cannot be avoided if we are to engage with grades at all. However, we can redesign grading practices in IST-related math courses to better serve our learning objectives and send a clear message to students about the behaviors and skills we value. Specifically, traditional grading based on homework correctness, timed exams, etc., sends the message to students that having the right answer or, worse, being able to parrot the right answer quickly is the goal.

Certainly, there is a body of work we can lean on in this regard. Research in education has much to stay about the alignment between grades and standards to promote learning (O'connor, 2017). Future work should explore grading strategies that support the objectives I have laid out.

Conclusion

Looking back, many of the key moments I can pinpoint on my path to a PhD in math and a career in IST research involved fun and exploration, free from concerns about why I needed to know it or why it mattered in the real world and free from concerns about evaluation. I loved playing with the abacus my parents gave me as child and playing Math Blaster! on our first family computer in the early 1990s.

In high school, I had the opportunity to explore some wonderful college math courses at a nearby university. I could pick courses I was interested in and attend them, but I was not expected to turn in assignments or take exams. Because there was no evaluation and no pressure, I selected topics that really interested me despite how daunting they seemed. I took a class in string theory one semester. I understood so little, but nevertheless, I remember feeling a rush just sitting surrounded by those big ideas. A young woman, I assume a graduate student, taught the class. I was mesmerized by her.

I had a similar experience in 2004. I was a junior in college. Manindra Agrawal spoke to a packed audience. He and collaborators had recently published the first polynomial-time primality test (Agrawal et al., 2004). With a few classmates, I took a NJ Transit train from New York City to Princeton to see him speak. I could not understand the details of the proof, but I felt the gravity of the moment. I knew that I was witnessing history; more specifically, I saw that math is alive, it is dynamic, and it precedes technology rather than responds to it. Mathematics is almost never motivated by application. By the time you know what you "need it" for, it must be there. That is the business of mathematics.

Among these formative experiences, there are common threads: curiosity; freedom from evaluation; untethering from real-world application; self-motivation; self-confidence; and self-esteem. When we develop information scientists, data scientists, and AI and technology researchers, I argue, we need to teach them mathematics the same way we teach mathematicians—not from an origin of application but treating application as a possible or plausible destination. This is because, like mathematicians, we are laying groundwork for a world of possibility.

References

Agrawal, M., Kayal, N., & Saxena, N. (2004). PRIMES is in P. *Annals of Mathematics*, 781–793.

Ascher, M. (1990) A river-crossing problem in cross-cultural perspective. *Mathematics Magazine, 63*(1), 26–29.

Ashcraft, M. H. (2002). Math anxiety: Personal, educational, and cognitive consequences. *Current Directions in Psychological Science, 11*(5), 181–185. https://doi.org/10.1111/1467-8721.00196

Bogomolny, A. (2024). Retrieved from https://www.cut-the-knot.org/

Cubitt, T., Perez-Garcia, D. & Wolf, M. (2015) Undecidability of the spectral gap. *Nature, 528*, 207–211.

Dowker, A., Sarkar, A., & Looi, C. Y. (2016). Mathematics anxiety: What have we learned in 60 years?. *Frontiers in Psychology, 7*, 508.

Dweck, C. (2016). What having a "growth mindset" actually means. *Harvard Business Review, 13*(2), 2–5.

Gill, R. (2010). Monty hall problem. *International Encyclopaedia of Statistical Science*, 858–863.

Gonthier, G. (2008) Formal proof–the four-color theorem. *Notices of the AMS 55*(11), 1382–1393.

Goodfellow, I., Bengio, Y., & Courville, A. (2016). *Deep learning*. MIT Press.

Hofstadter, D. R. (1999). *Gödel, Escher, Bach: an eternal golden braid*. Basic Books.

Mendick, H. (2005). A beautiful myth? The gendering of being/doing 'good at maths'. *Gender and Education, 17*(2), 203–219.

Middleton, J. A., & Spanias, P. A. (1999). Motivation for achievement in mathematics: Findings, generalizations, and criticisms of the research. *Journal for Research in Mathematics Education, 30*(1), 65–88.

NIST Announces First Four Quantum-Resistant Cryptographic Algorithms. (2022). Retrieved from https://www.nist.gov/news-events/news/2022/07/nist-announces-first-four-quantum-resistant-cryptographic-algorithms

Nosek, B. A., Banaji, M. R., & Greenwald, A. G. (2002). Math = male, me = female, therefore math me. *Journal of Personality and Social Psychology, 83*(1), 44–59.

O'connor, K. (2017). *How to grade for learning: Linking grades to standards*. Corwin Press.

(2022). Retrieved from https://www.claymath.org/millennium-problems/

Ribenboim, P. (1979) *13 lectures on Fermat's last theorem*. Springer Science & Business Media.

Tobias, S. (1993). *Overcoming math anxiety*. WW Norton & Company.

Wiles, A. (1995). Modular elliptic curves and Fermat's last theorem. *Annals of Mathematics, 141*(3), 443–551.

Yu, X., Zhou, H., Sheng, P., Ren, B., Wang, Y., Wang, H., & Zhou, X. (2024). Math anxiety is more closely associated with math performance in female students than in male students. *Current Psychology: A Journal for Diverse Perspectives on Diverse Psychological Issues, 43*(2), 1381–1394. https://doi.org/10.1007/s12144-023-04349-y

Professor in a Strange Land: An Expat Metaphor for Classroom Discussions

Shomir Wilson

Abstract Classroom discussions provide many benefits to students, including practice with verbal skills, enhanced critical thinking skills, an active learning framework, and greater intellectual stimulation than an unidirectional lecture format. Although discussions' suitability for learning objectives varies (e.g., for some mathematical skills, a different active learning format could be more suitable), many courses cover material that classroom discussions can enhance. However, successful classroom discussions present obstacles for instructors, including a lack of confidence in the format's success or an inability to relate to students' perspectives. I present an "expat metaphor" for how instructors can internally relate to students during class discussions. Rather than avoiding the gap in cultural referents between the instructor and students, the metaphor exploits it toward a way of thinking about classroom discussions that is egalitarian while maintaining the instructor's authority. I describe how I run discussions in my courses, link the metaphor to them, and explain how the metaphor frames planning and moderation of discussions.

Keywords Discussion-based learning · Active learning · Instructional strategy

Introduction

Classroom discussions help students develop verbal skills, self-evaluate their understanding of course material, and reason through problems in an interactive setting (Zare & Othman, 2015; Murphy et al., 2016). They further encourage student engagement by setting aside the unidirectional communication model of lectures (Akçayır & Akçayır, 2018). In an informatics context, they provide a teaching method for content with a strong verbal component and, more broadly, support a

S. Wilson (✉)
Pennsylvania State University, University Park, PA, USA
e-mail: shomir@psu.edu

J. M. Carroll (ed.), *Innovative Practices in Teaching Information Sciences and Technology*, https://doi.org/10.1007/978-3-031-61290-9_3

liberal education-like engagement for students with a wide variety of interests (Berg & Lee, 2016). From the instructor's perspective, they can be intellectually stimulating to prepare for and engaging to moderate. However, in spite of the appeal of classroom discussions, they remain difficult for some faculty to plan and to implement with confidence. Recognizing this obstacle and finding ways to reduce it supports a trend in tertiary education away from lectures and toward active learning.

In this chapter, I describe an "expat" metaphor for relating to students, in which the instructor is an expatriate living in a society where the students are indigenous; this metaphor frames classroom discussions in a way that may reduce the perceived difficulty of using them as a teaching method. The metaphor has no explicit presence in my interactions with students, but it informs how I guide some of those interactions. At the core of the metaphor is a recognition that the instructor is surrounded by students whose connections to contemporary sociotechnical issues are different and often more immediate. These differences arise from intuitive sources: privileges of age, experience, and relative wealth; the emphasis on youth culture in society and economy; and certain goals implicit in being a student. Rather than treating these differences as obstacles, the expat metaphor exploits them as a source for qualitative contrast, with an emphasis on students' more relevant (and *de facto* more numerous) perspectives. The metaphor also frames important similarities and differences between instructor and students: although both parties' experiences and cultural referents deserve respect, the students are effectively indigenous to the context. For the instructor, the recognition is a way of thinking about how to respect students and a motivation for interest in their concerns.

I then describe experiences with classroom discussions in two undergraduate courses. The first is an honors introductory-level course that attracts students from a variety of majors, and the second is a course on consumer privacy that cybersecurity majors typically take in their fourth year. Sections of the honors course typically consist of 10–20 students, while sections of the privacy course consist of 60–80 students. In all three courses, I divide classroom time between lectures and discussion, but the procedures for discussions vary by class size, with my instructions being more flexible for the smaller courses and more formal for the larger course. Early in the semester, I ask students to reflect on what makes a conversation work well, and I refer to metacognition periodically during the semester. The classroom discussions encourage students to reconcile information in the assigned readings with their own experiences, to synthesize a stronger understanding of course content. The expat metaphor provides a framework for students and the instructor to have rich discussions on technology and society, often leading to novel conclusions in place of predefined ones. I would argue that the authenticity of this synthesis (i.e., both parties learning something from the discussion) enhances the students' learning experience.

The Value of Classroom Discussions

Classroom discussions are a valuable part of pedagogy. By encouraging students to ask questions, self-assess, and selectively improve their understanding of a topic, classroom discussions belong among a set of teaching methods known as *active learning*. They further encourage students to think critically and to develop confidence verbally working through problems with their peers. Verbal skills that students gain from classroom discussions will continue to serve them in collaborative environments in their future careers. Additionally, the interactive nature of discussions can increase student engagement and enjoyment of classroom time, leading to greater retention of course material.

Within informatics, classroom discussions juxtapose with other in-class teaching activities, including lectures and hands-on learning. None of these activities is suitable to replace all of them, and each has qualities that justify its inclusion in a typical course. Programming skills, for example, are sometimes easier to learn with classroom time to work on exercises in groups with an instructor close at hand to provide assistance. Students may want an instructor to explain a complex topic first (i.e., through lecturing) before they engage with it directly. In contrast, to explore conceptual problems or develop skills answering questions that involve value judgemnts, discussions provide students an opportunity to reason through finding an answer with the help of their peers and the guidance of an instructor.

Students at different stages of their tertiary education also may benefit in different ways from classroom discussions. First-year undergraduates' early impressions of college are likely to have an outsized impact on their experience going forward. Being shown, even if implicitly, that their ideas have value is likely to support their confidence in their abilities to learn. "Big picture" discussions of their discipline also lend to students' abilities to talk about it and to articulate their interests. Further along, students in their final year of college (as well as graduate students) have reached a point in their education where summative reflection and reasoning about their discipline are meaningful and can bridge their entry into their professions. There is an esthetic symmetry between discussions at these two stages: early discussions help to illuminate the landscape that students will learn about, and later discussions crystallize the memory of it as students transition into their careers.

The instructor also may benefit, acknowledging that teaching strengths and dispositions may vary among faculty. For some, preparation for a lecture, including planning out time usage and creating aids such as slides, is more time-consuming and arduous than preparation for a discussion. The converse may be true for others, as a good classroom discussion requires some preparation: selection of assigned reading, a brief mini-lecture to frame the topic and the conversations to follow, thoughtfully chosen questions to guide discussion, and groundwork laid earlier in the semester to help students feel comfortable sharing their thoughts. Even then, more is left to chance for the discussion than the lecture: the class could be unresponsive to the topic, individual students could be unhelpful (e.g., dominating the discussion or leading it away from the preferred topic), or the social chemistry of a specific class could be problematic. An instructor's risk tolerance for these scenarios may depend on a variety of variables, including personal preference and institutional climate.

However, the distribution of faculty (i.e., across comfort levels for classroom discussions) is neither binary nor static. It seems possible that some instructors avoid having classroom discussions but would be open to trying them. Some of the obstacles to trying them are a fear of a conversation becoming out of control, inability to judge what will provoke students to participate, or low confidence in making effective use of the available time. For that audience in particular, I provide the expat metaphor as a way to think about these obstacles and overcome them.

The Metaphor, Referents, and the Classroom Setting

Overall, the idea of the expat metaphor is that the students are native to the sociotechnical context of a discussion topic, and the instructor is an expatriate in their society. While remembering important boundaries and the necessity of classroom authority, the instructor can provoke valuable discussion and reflection among the students by asking questions about how the course material relates to the students' cultural and technical frame of reference. The metaphor reinforces the notion that students' knowledge perspectives are relevant to learning and worthy of respect while encouraging students to explore and deconstruct those perspectives. This metaphor might be primarily relevant for topics that merit discussion about social issues, but even in many technical courses, I would suggest that some discussion of social issues (especially ethics and bias) is appropriate.

Cultural and sociotechnical referents (simply *referents* below) are an important for the metaphor. These are, broadly speaking, items that people widely can discuss. They include creative artifacts such as TV shows, movies, books, online videos, and memes, but they also include technologies that have impacted culture (smartphones, streaming video, ride-sharing services, etc.) and other concepts or attitudes that many people are familiar with (the gig economy, fear of missing out, surveillance, sharenting, etc.). Equal understanding of them is unlikely to be universal, and instead, it may be segmented into cohorts of people with shared cultural referents.

Having experienced being an expat helps understand the metaphor, although that experience is not essential. I draw upon a year I spent as a postdoc living in Scotland and respective periods of ten weeks in Australia and Singapore. The overlap between my cultural referents (having lived most of my life in the United States) and those places was partial, with some shared referents but many different ones. A (mostly) shared language and some common ground provided a basis for me to understand the cultures I lived in. People I met in work and in social contexts expressed some interest in my US-centric reference point, but it typically led to more engaging conversations if I asked questions about the cultural setting I was in. That interest both established rapport and increased my ability to understand the setting.

Many of the above observations have analogs for how a faculty member relates to their students' cultural and sociotechnical reference points. Crucially, the difference in reference points depends on normative assumptions about age and relative wealth: a typical college instructor is substantially older than their students, leading to a variety of differences in experiences with culture and technology, and the instructor may have greater privileges of stability in their career and finances. The instructor's past is likely to have cultural and technological referents that they relate to most comfortably, but using those referents to communicate with younger people is less likely to produce the desired results. The instructor may approximately understand the students' referents, but not with their native fluency. The instructor may flaunt their lack of understanding for humorous effect, but the amusement is likely to be short-lived.

Instead, an effective approach for the instructor to generate discussion is to express interest in the referents that they are surrounded by, i.e., the reference points in the setting of the present, which the students are native to. Naturally, the conversation should be directed toward referents that are relevant for the course material and those that the instructor is familiar enough with to guide the discussion. Identifying those referents is the familiar task of finding connections between contemporary society and course content, but the instructor's role is less didactic than it is inquisitive: *these are contemporary issues, from your time, and I'm curious what you think of them.* While using contemporary examples is itself unremarkable in teaching, the expat metaphor reinforces respect for students' perspectives and, in a limited but potentially empowering way, recognizes them as near peers to the instructor in a discussion.

The metaphor has limitations and caveats. As always the case with classroom discussions, the instructor ultimately relies on the students to generate respectful, productive discussion. It is possible to prime students to have good discussions with appropriate guidance, and I describe some of my approaches to this in the next section, but there are limits to an instructor's powers. Additionally, it is possible for students to reach conclusions that conflict with the learning objectives. The metaphor makes the instructor's role delicate: they must respect students' expressed thoughts while realigning them with the objectives. Finally, although the metaphor creates an egalitarian space for evaluating ideas, outside that space it is appropriate for the instructor to maintain their authority and their boundaries with the students.

The Courses

The expat metaphor informs my teaching style for several courses, including seminars for first-year undergraduates, upper-level undergraduate electives, and graduate courses. I highlight two courses in this section as examples for comparison and contrast.

Course #1: Honors Information, People, and Technology

This course introduces students to the landscape of information sciences and technology, but it serves additional roles for the institution. As an honors course, it covers material beyond the standard sections and provides students with honors credit. Instructors' choices for the additional materials vary, and in my sections, I include content on technology ethics and critical thinking about the impacts of computing technology on society. Approximately, a third of the course is a nontechnical overview of basic information science and information technology (IS/IT) topics, such as hardware, software, data and databases, networking and communication, information systems security, and information systems development. I use an open textbook by Bourgeois et al. (2019) to guide this learning. Another third of the course is an exploration of technology ethics, guided with readings from Vallor's *Technology and the Virtues* (Vallor, 2016). The final third satisfies the course's institutional role as a *first-year seminar*: a course that introduces students to the expectations of college and resources to help them succeed. Throughout the semester, guest instructors present and lead activities around career preparation, teamwork, majors and options within the college, academic integrity, and library skills.

Structure for In-Class Discussion

Enrollment in this course tends to 10–20 students. Within this size range, I find it feasible to have one collective discussion (i.e., instead of smaller breakout discussions), though moderation remains necessary to make sure all students participate and the discussion remains productive.

One moderation strategy is the use of poker chips as tokens to track discussion contributions. I pass around a bag of these at the beginning of class and ask students to take three, one of each color: blue, red, and white. I instruct students to turn in their blue chip when they make their first discussion contribution and then their red chip for the second contribution. Often, I ask them to wait until everyone has contributed at least once before allowing a third contribution (thereby turning in their red chip), and sometimes when all blue chips have been turned in, I open the floor to unlimited repeat contributions. By asking students to show their chips on their desks, I can assess visually who has yet to speak and how evenly discussion contributions are distributed. Toward the end of class, if some students still have not contributed, I tell them they will be my "panel" (with mild humorous theatrics, to help them relax) and ask them to share their thoughts on a simple open-ended question. This ensures that all students have an opportunity to receive participation credit. Students turn in all their remaining chips at the end of the class session.

As a one-time experiment, I asked students for input on other ways we could use the colors of the chips. Briefly, I experimented with blue for leading contributions, red for replies to others' contributions, and white for a Facebook-inspired *like* of others' contributions. After trying those options, I asked students whether they enhanced the discussion, and the verdict was negative. Accordingly, I used the simpler system described above.

The role that discussions play in class sessions depends on the material we are covering. During sessions in the IS/IT portion of the semester, I begin class with a lecture on the textbook material, and approximately halfway through the class time, we switch to discussing a news article that is part of the assigned reading. Later, during sessions in the technology ethics portion of the semester, we begin immediately with discussion. By beginning with the half-session discussion format and then switching to the full-session format later in the semester, I allow students (and myself) some time to assess how these discussions work with the specific composition of the class before extending their duration. For both formats, I begin the discussion by speaking for a few minutes with my thoughts on the reading. I segue into focusing on class contributions with an open-ended question, and I keep several more questions available in my notes to guide the session and revive discussion if it reaches a lull.

Reflections

The expat framing provides guidance for the kinds of discussion questions that work well in this course and how to respond to students' answers. The observations I make below might not be unique individually, but the metaphor bundles them into a cohesive form.

I typically focus the discussion on open-ended questions that I also (i.e., in addition to the students) find intellectually stimulating. Factual recall questions tend not to work well for multiple reasons: they create tension and risk for the students (e.g., being afraid to answer because of the risk of being incorrect), they lead students away from being able to reason through material, and they represent a mode of assessment that can be done more efficiently through exams and take-home assignments. I sometimes emphasize the open-endedness of the questions and my mutual curiosity, to add an egalitarian tone to the discussion. Examples of questions like these include:

1. In response to an article about employees feeling pressured to overwork at Rockstar Games (Inside Rockstar Games' Culture of Crunch, 2018), how should a manager balance the interests of a company with the well-being of its employees?
2. In response to a claim in assigned reading that "every company needs to become a software company" (Patel, 2016), what does it mean to act like a software company? Ostensibly, this is discussed in the reading, but many answers exist.
3. In response to an assigned reading about predictive analytics for child protective services (Hurley, 2018), what unintended consequences come with a pivot from focusing on children known to be at risk to those predicted to be at risk? In

general, why is it difficult to get the benefits of predictive analytics for social good without drawbacks?
4. In response to an article about Stanislav Petrov (2013), whose distrust of an early-warning missile detection system may have prevented a nuclear war, when is it appropriate to doubt warnings from complex technical systems? How do our value systems function when time is limited and mistakes are disastrous?

These questions illustrate a variety of cultural referents that may be nearer or further from students' native experiences. For the first question, Rockstar Games is responsible for many popular video games that current college students are likely to have played. The "software company" mentality in the second question is pervasive to the extent that younger people may take it for granted, a correlary of the article's thesis. The third question addresses the role of analytics in society, and like the second, its pervasiveness is easily taken for granted by their age demographic. Conceptually, the fourth question is distant from the direct experiences of both the students and the instructor.

In general, I observe greater engagement with questions that are closer to students' native frame of reference. The introductory part of a class session, whether a half-lecture or a few minutes of my thoughts on the subject matter, provides additional framing to reposition those referents into the topics of the course, enabling the learning goals and allowing me to guide the conversation. I sometimes leverage my professed curiosity about the students' frame of reference to encourage them to explain referents that their peers could be unevenly unfamiliar with, such as popular memes or movies. This further helps demonstrate for students the value of their perspectives to understanding course topics.

Course #2: Integration of Privacy and Security

This course is a nontechnical introduction to computing privacy, with a focus on consumer privacy. It is a fourth-year undergraduate course that mainly enrolls students in the college's cybersecurity analytics and operations major. Students in this major have seen a plethora of content in prior courses about security, and this course serves as a counterweight toward respecting users' autonomy and preferences. Course topics include foundational theories of privacy, privacy law, the basics of differential privacy and encryption, dark patterns, ubiquitous computing, and intersections between privacy and discrimination. I use a published collection, *The Cambridge Handbook of Consumer Privacy* (Selinger et al., 2018) for readings and inspiration for many class session topics. Graded assessments consist of reaction papers, participation in in-class discussions, and a term project. The term project is a group activity in two stages: a presentation of a notable consumer privacy failure and a presentation of a plan that would have avoided or mitigated the failure.

Structure for In-Class Discussion

Enrollment in the course tends to be 60–80 students. Within this size range, only a limited amount of collective discussion is feasible; I often ask questions during lectures, but it is difficult to ensure that everyone participates and benefits from the interaction. Instead, I make frequent use of breakout discussions to distribute and parallelize participation. Toward the beginning of the semester, I explain that discussion will play a large role in the class time, and I ask students for input on what makes conversations work well. This question is explicitly for a broad scope: I encourage answers for the classroom setting and for less formal settings (e.g., the setting of a group of friends), highlighting similarities and differences. I collect the answers and review them with students, and following that, I show them the answers from the prior semester that I taught the course. This connection to others' prior efforts reinforces the validity of their work and the importance of the exercise.

Toward the end of most class sessions, I present students with one or two discussion questions. These are open-ended in a way that is similar to the discussion questions for Course #1 above: I tend to ask questions that are intellectually stimulating both for the students and for me. I ask students to get into groups of 3–4, emphasizing that two is too few and five is too many. Groups write their names and a brief summary of their answers on a sheet of paper, which they turn in to the teaching assistant (TA) at the end of class to receive participation credit. I instruct the TA to grade for completion rather than answer content. While students talk in their groups, the TA and I circulate around the room to ensure students remain attentive to the task, to answer questions, and to briefly participate in their discussions.

After a pre-announced time period (typically five to seven minutes), I ask groups to reconvene for a collective conversation. I begin this conversation by asking for "priority answers," those that groups are particularly proud of or especially want feedback for. After that, I may ask individual groups to share their answer, highlight groups that I spoke with while circulating and ask them to speak, or ask for answers that disagree with those that have already been spoken. These techniques tend to be sufficient to seed a self-sustaining discussion, with students pivoting from speaking in response to me to volunteering to respond to their peers' contributions.

Reflections

The questions that I ask in Course #2 retain an essential open-endedness, although they sometimes incorporate basic problem-solving. Consumer privacy is a fertile topic for questions that involve tradeoffs between respect for individuals, protection of individuals, risk minimization for companies, fidelity to best practices, innovation, familiarity, usability, and fairness. Some example questions include:

1. In 2015, the education testing company Pearson monitored Twitter for public posts that indicated students might be leaking questions from its PARCC standardized tests. In at least one case, Pearson contacted the school district of a

student who posted test content, and the student was disciplined (Singer, 2015). Was this appropriate for Pearson to do?

2. Are conversational technologies like Alexa and Siri a problem for children's development of social skills? If yes, how should parents or society respond to the problem?
3. Do different types of smartphone data (e.g., location, text messages, photos, etc.) deserve different levels of specificity when explaining data practices to consumers?
4. A store decides to use small Bluetooth beacons (i.e., small devices that sense when Bluetooth-enabled mobile devices are nearby) to keep track of customers' locations as they walk through the store, exploiting the fact that many customers have Bluetooth-capable smartphones or smartwatches. What is (a) an outrageous way to notify customers about this data collection and (b) a way to provide notice that will not discourage any customers from entering the store?

The expat metaphor is most salient for the first question: when I took standardized tests as a student, social media was in its infancy, and it was a negligible factor for testing integrity. I can infer and predict what concerns exist, but it is meaningful for me to disclose that students' experiences are more immediate than mine and their contributions to the topic are meaningful. The second question asks students to reflect on a time in their life that was more recent for them than it is for me, and to assume some distance from it, assessing (regardless of their personal experience) whether childhood use of a certain technology poses problems. Again, this is a combination of a technology and a stage of life that they are more familiar with than me, though I can reason about their claims. The third question addresses a technology that we (i.e., the students and me) use with equivalent fluidity but possibly different attitudes toward that use. Finally, the fourth question opens the door to differences in attitudes toward surveillance, with an invitation for both outrageous, amusing answers and thoughtful ones.

I again observe greater engagement with the questions that are closer to the students' native frame of reference. The procedure of small group discussions followed by a large discussion appears to help students speak candidly, without the scrutiny of a large audience, and vet their answers so that they can speak more confidently when the time comes to share with the entire class. Meanwhile, the expat framing is one way for me to think through respecting student answers while reasoning about them and guiding the discussion toward learning goals.

Conclusion

Student engagement in classroom discussions is valuable and productive—and sometimes scarce or fragile. Questions that students do not relate to or understand can produce silences that even a patient instructor cannot bear to wait out. Instructor responses to students' discussion contributions must be tuned to avoid being overbearing, patronizing, or dismissive.

A completely egalitarian discussion is, in a *de facto* sense, impossible in an instructor-led classroom. However, the expat metaphor provides some scaffolding for the instructor to organize their thoughts on how they and the students, respectively, relate to the course material. That organization, as I envision it in this chapter, can provide greater confidence and intentionality to the instructor, potentially contributing to successful learning outcomes.

References

Akçayır, G., & Akçayır, M. (2018). The flipped classroom: A review of its advantages and challenges. *Computers & Education, 126*, 334–345.

Berg, D. R., & Lee, T. (2016) Incorporation of liberal education into the engineering curriculum at a polytechnic. In *2016 ASEE Annual Conference & Exposition*.

Bourgeois, D. T., Smith, J. L., Wang, S., & Mortati, J. (2019). *Information systems for business and beyond*. Saylor Academy.

Hurley, D. (2018). *Can an Algorithm Tell When Kids Are in Danger?* The New York Times. https://www.nytimes.com/2018/01/02/magazine/can-an-algorithm-tell-when-kids-are-in-danger.html

Inside Rockstar Games' Culture of Crunch (2018). https://kotaku.com/inside-rockstar-games-culture-of-crunch-1829936466

Murphy, P. K., Firetto, C. M., Wei, L., Li, M., & Croninger, R. M. (2016). What really works: Optimizing classroom discussions to promote comprehension and critical-analytic thinking. *Policy Insights from the Behavioral and Brain Sciences, 3*(1), 27–35.

Patel, J. (2016). *Software is still eating the world*. https://techcrunch.com/2016/06/07/software-is-eating-the-world-5-years-later/

Selinger, E., Polonetsky, J., & Tene, O. (2018). *The Cambridge handbook of consumer privacy*. Cambridge University Press.

Singer, N. (2015). *Pearson under fire for monitoring students' twitter posts*. https://archive.nytimes.com/bits.blogs.nytimes.com/2015/03/17/pearson-under-fire-for-monitoring-students-twitter-posts/. Cad: 0 Section: Technology.

Stanislav Petrov (2013). The man who may have saved the world . *BBC News*. https://www.bbc.com/news/world-europe-24280831

Vallor, S. (2016). *Technology and the virtues: A philosophical guide to a future worth wanting*. Oxford University Press.

Zare, P., & Othman, M. (2015). Students' perceptions toward using classroom debate to develop critical thinking and oral communication ability. *Asian Social Science, 11*(9), 158.

When Less Is More: A Java Introductory Course

Fred Fonseca

Abstract Learning a programming language can feel overwhelming for students. The excess of details can stand in the way of a more holistic grasp of the basic structures and concepts of a language. In this chapter, I describe an attempt at using minimal concepts and simplifications to teach a new programming language. This is told from a personal point of view using my own experiences designing and delivering an introductory Java programming course.

Keywords Problem-based learning · Teaching programing languages · Minimalism

Introduction

Learning a new programming language can feel overwhelming for some students. This is a feeling I had as a student when I had to learn a new language. This is also a feeling I have when I see programming courses that try to teach everything to someone who knows little.

This chapter describes an innovative approach to teaching Java. It is a programming course that comes after an introductory programming course. In the prerequisite course, students learn basic statements such as variable declarations and if statements either in C++ or Java.

In this Java course, we opt to have a series of assignments with each focusing on a basic concept. There are three series of labs/assignments focusing on the large concepts in Java: classes, inheritance, and arrays. The labs are sequential. Each lab develops one step toward the final lab. The first lab is quite simple, and the last is complex for an introductory course in Java.

F. Fonseca (✉)
College of Information Sciences and Technology, Pennsylvania State University, University Park, PA, USA
e-mail: fredfonseca@ist.psu.edu

J. M. Carroll (ed.), *Innovative Practices in Teaching Information Sciences and Technology*, https://doi.org/10.1007/978-3-031-61290-9_4

Two things make this approach different from traditional approaches to teaching introductory programming. First, I employ a minimalist perspective in teaching programming. While other courses often emphasize a heavy theoretical approach, or strict standards on coding style, or even a detailed discussion on how a Java IDE (integrated development environments) works, I opt to discern the most basic concepts in Java and develop each lab accordingly.

Second, the approach presents a difficult challenge and a plan to address it. This is where the sequence of the labs plays a key role. The first lesson in the course begins by presenting a fully developed Java application, which is the final lab, on the very first day. Over the rest of the course, I share with the students a path to get there and let them know that it can be done if they stick to the week-by-week plan.

Motivation

I wonder how many instructors shape their courses based on their own learning styles. I always had problems grasping a new programming language. Every time I was learning a new programming language, I always felt a huge barrier between myself and my first running application. Depending on the language and the companion program interface, this can be a difficult first step to take. Even a "Hello World" program can take some time to successfully run.

A Hello World program is a simple application often used as the first program in a programming course. It has the drawback of being too simplistic; it just displays one line saying "Hello, world." It has the advantage of being simple and doable. One of the problems is that it is often hidden behind a sometimes complex programming interface (IDE https://en.wikipedia.org/wiki/Integrated_development_environment). The interface might be the first problem the student encounters. Although important, I will not address the IDE complexity problem here and will instead focus on how to teach the basic Java concepts in an introductory course.

One of my goals in the course was to enable students as soon as possible to have a small and meaningful application, more than a Hello World or a slightly improved version of it, in Java. This would be not only the entryway to the language but also the way to keep working on all the subsequent concepts and programming structures.

The main guideline is then to keep it simple and add incremental small steps. The way to implement this was a series of labs (small programming applications).

Another important motivation for me is for students to learn with their hands. Once I told a student that this or that statement should be coming from their hands, not from their minds. What I meant by this was that to master a language, some basic concepts and structures just have to come as you type. This mastery comes from repeatedly practicing the coding language.

The act of programming, writing an application, debugging it, and running it has some level of engagement that is fundamental to learning. Letting students try different routes to solve a problem is very important. Even when they are going to a dead end. They need to do that to see how and why they got there. And from there they can develop their own way to use the language.

The Course Structure

I use two variations in this introductory Java course. In one version, I have 7 labs and a big project that takes 7 weeks to complete. In another, I have 13 labs and no final project. The final project and achievement is the last completed lab. I will be describing the 13-lab course here, although I may reference the 7-lab course as well to illustrate some points.

I use a weekly structure. The course might be on a 2- or 3-day-a-week schedule. There is one concept every week. On the first day of the week, I give a brief introduction to the topic. This is followed by an example that is created live in the classroom with the students following and typing on their own computers. After that, the students work in a small programming project, a lab that is like the one they have to deliver at the end of the week.

I always start a lesson by showing the students the last lab. I run the lab and tell them that it is a complex application. It has seven classes (the first lab has only two). The purpose of showing a seemingly unachievable objective in the first minutes of the first day of the semester is to try to convince the students to engage in the step-by-step approach. Right after showing the final product of the semester, I show them the lab sequence including the first one which then seems very simple and achievable.

The Labs

The labs build on each other. This means that the last (13th) lab has components of—and needs—every single lab before it. Each lab adds some functionality to the final application.

Although this approach makes it difficult for a student to tune in to the course in the middle of the semester, it embeds the learning path to get where they need to be. In another article (Fonseca & Spence, 2014), I described what I called the Kid approach to teaching programming languages. I was teaching this same course and after a series of seven labs, a student got stuck when I asked them to develop a full application, a simple game that they would design and program themselves. They came with an often-asked question: "I am lost, I don't know anything, I don't know how to do this." I replied: "How do you want to start? What is the first thing you want to build?" The student said: "I don't know how to pass information from one class to another." Then I replied, "This is Lab 4 and if you don't know how to do it, go back to Lab 3" and so forth.

The inspiration for the name of the method came from the Karate Kid movie moment when Daniel, having to face a real opponent, tells the master that he doesn't know anything, he doesn't know how to practice Karate. Mr. Miyagi, the master, then shows Daniel that each movement he needs he learned doing apparently unrelated tasks. Mr. Miyagi then attacks Daniel. Daniel must defend himself. Mr. Miyagi then tells him "Show me paint the fence. Up, down." For another important movement, Mr. Miyagi tells Daniel: "Show me sand the floor." Building on what he had learned before, he realized he could do Karate. This is the same approach with the labs—though it may be difficult to see at first, all the labs are interrelated and building to achieve the ultimate goal of learning Java.

Basic Concepts in Java

Java is an object-oriented language. The concept of objects, which are data and code together, is a fundamental construct. Java is "class-based, meaning that objects are instances of classes, which also determine their types" (https://en.wikipedia.org/wiki/Object-oriented_programming).

So, here in this simple definition, we have classes, objects, and types, the fundamental constructs of Java. All of them work together and can have different levels of complexity. The key is to find the most basic example and lab that can introduce the three concepts and let the students have a hands-on approach to learning.

Minimalism in Practice

Many programming courses start with a version of Hello World. So, here I am overriding what is considered the most basic programming application in an article that is supposed to argue in favor of simplicity. My own version of Hello World is a bit more complex but also much more meaningful.

The First Day: Creating a Basic Class

Instead of using the overly simplistic Hello World approach, I opted to start by creating a class. A class in Java can be a complex construct. So, I start by creating a very basic one. It is so basic that it is actually incomplete. The full structure of a complete class will be achieved by the fifth lab in this initial sequence.

This basic class contains only the attributes, the data descriptors that are used as "class variables." They are variables that are available for the whole class. The class makes use of the default Java constructor. Java embeds a constructor in a class if the programmer doesn't provide one. This is a good topic to be discussed as an add-on in the third week, constructors.

Here we have the application creating an object. The *new* statement, the Java keyword to initialize an object, is used and discussed. Also, the *dot* operator is introduced to retrieve data from the object and then display it.

Finally, students are asked to use *if* statements, something they should have seen in a prerequisite course. The *if* statements—used in the first programming exercise—use objects and the *dot* operator to display some data.

At the end of this first day/lesson, the student will have touched the three most important concepts in an object-oriented language: classes, objects, and object-oriented programming.

I call this approach minimalist because you can't have *less* than that on the first day. You could have more which is a temptation that every instructor has, trying to teach too much. It is also minimalistic because it tries to avoid unnecessary structures. Yes, there is much more to a class in Java, but these are topics for the next 4 days/lessons.

The Second Day: Making the Java Class More Complex

Building on the first-day lab, the second lab adds methods to the class. Methods are a set of statements in a clearly defined region of the class code. They often do some calculations or data processing. They can return an object or a value at the end or just complete a task without returning anything.

The second lab asks students to create methods. One is a basic method in Java, the *toString* method. It returns a string representing all the data in a class plus some labels identifying the data.

Another method in the lab asks students to bring the code that was used outside this specific class, in the main application, and place it in a method. This leads to a discussion of what belongs to this class and what belongs to other classes.

The class they are building is becoming more complex. It now has attributes from lesson 1 and methods from lesson 2. The next step is to add a special method to the class: the *constructor*.

The Third Day: Constructors

A constructor is a special method in Java used to initialize an object. So far, students have been creating objects using the default constructor provided by Java. Then they assign values to each attribute individually from the application. The constructor they will create in this lesson will do this in a more streamlined way.

Two concepts from the last lesson will be rehashed here. The first is object initialization. The students did it in the first week. Now, with constructors, they will do it in a more sophisticated way. The second concept, receiving and passing information between classes through input parameters, is also used more in depth here. The way constructors are used, receiving information from outside as input parameters, will also reinforce the *methods* concept from the last lesson.

The Fourth Day: Complex Types

This lesson reinforces the concept of *types* which is fundamental in Java. Students have already been using types since Day 1. In Java, every time you define a variable (or object), you need to specify its type. For example,

```
int counter;
String name;
Double gpa;
```

In the above declarations, "int," "String," and "Double" are types. They define what kind of data you can store in a variable.

Students have already learned how to create a basic class, for instance, *Address*.

```
public class Address
{
    int number;
    String street;
    String city;
    int zip;
//.....
}
```

Now, combining we can make a class they are already familiar with, Person, more complex. We can add a new attribute, *homeAddress*, using the type *Address*.

```
public class Person
{

    String name;
    int weight;
    String hometown;
    String highschool;
    Address homeAddress;
    //.....
}
```

This is a subtle but powerful change that adds some complexity to the use of concepts they have used in the first three lessons. They know what types are. They know what classes are. But they don't know yet that classes can be used as types. The great conclusion is that in Java classes can also be used as types.

The new attribute will carry a full class inside it. This has lots of consequences that students will need to deal with in this lesson. I often tell the students at the beginning of the class that there are no new *concepts* this week. It is just a different, more sophisticated, use of what they already know.

One challenge is understanding how classes (or objects) communicate with each other. Now they can use, or better *build*, a statement that will access the class *Person* and then the *Address* inside *Person*.

If you have an object of the type *Person*:

Person p1;
you can, for instance, display the city where the Person lives.

```
System.out.println(  p1.homeAddress.city  );
```

We build this statement together step by step. This is done using Java's DOT operator, something they learned in lesson 1.

Starting with *p1*. We can see what is available in *Person*:

We picked *homeAddress*. Then, again, using the DOT operator, we move and see what is available inside *homeAddress*, which is an *Address*:

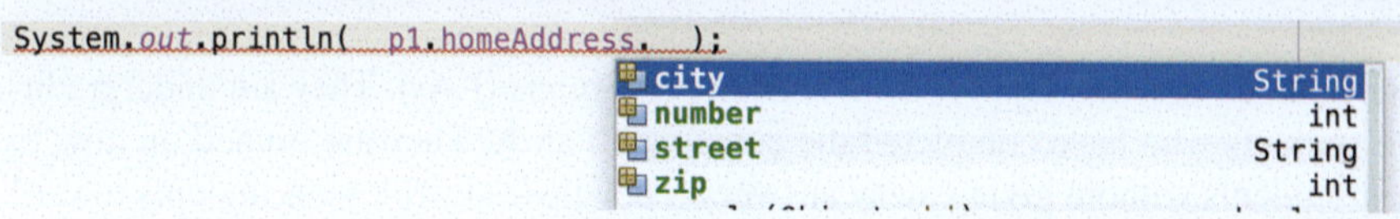

We picked *city*. With these two choices, we show how a class can be embedded inside another class and how this can be used. The rule can be applied for another class inside *Address* and so forth. We are using a simple statement to demonstrate a complex concept.

This lab also has another interesting rework of previous concepts. For example, in lesson 1, they created objects and used them outside a class. Now they are doing the same thing inside a class. An *Address* object is created inside the *Person* object.

The Fifth Day: A Complete Java Class

The big achievement in this lesson is that students will have a complete class built. It has the main components of a standard Java class. The topic is called *encapsulation* or *information hiding*. They are used to protect the data in a class and offer access for retrieval and updates in an organized way.

This lesson wraps up all the previous concepts, closing the unit called *Classes and Objects*. It also shows a class more similar to what students will find in textbooks and online examples.

The course continues in a similar fashion discussing another important concept in object orientation, *inheritance*. After that, the last unit in the course covers *arrays*.

The Classroom

Because I always teach in a large classroom with lots of computers, it is not unusual for students not registered in the course to step in and try to use the computers. Their reasoning is that this is not a course in session because, first, I can't see the instructor, and second, there is too much talking and discussing for this to be a class.

This is because I am always walking around and sitting with students discussing their problems. I also encourage students to discuss programming problems with their peers. It is made clear that they can discuss the problems, but they can't copy solutions or code. This creates an interactive and animated environment that simulates something like a start-up rather than a traditional classroom.

Classroom Support

The course often has support from *learning assistants* (LAs). They are undergraduate students who have completed the course with an A. They are trained on how to help students without giving away answers. They instead work with students toward solutions in a discussion manner.

The Atmosphere

All this creates a lively atmosphere. Some students might prefer a quieter atmosphere, and they are allowed to use headphones. When I need to discuss specific issues or just ask how the projects are going, I often have to take some of them out of *the zone*, in which they are highly focused on the code but listening to something else.

Can Minimalism Be Used in Advanced Courses?

There is one topic in the next Java course that fits very well with minimalism. MVC (model-view-controller) is a design pattern, a way to develop applications. It tries to separate three important elements in application design, the data and "intelligence," the display, all the output that the system produces, and, a third element in charge of making things happen, the controller.

In an MVC application, *Controller* calls *Model* asking for data, and sends the data to be displayed to *View*. *View* can also be the place for the user to interact with the interface and generate some requests for data. *Controller* becomes aware of the user interactions and responds by asking *Model* for some data and passes the data back to *View*.

In this course I use one Java statement to show the students how it synthesizes all principles of MVC:

```
view.basicDisplay(model.getData(0));
```

After presenting this statement, I erase it and start building it again this time with the students doing the same on their computers.

I tell the students that if we want to display something, then we need to use the resources available in the *View* module. I then start by typing *view* which is the object representing the *View* class. This triggers the IDE autocomplete menu with the possible options to complete the statement.

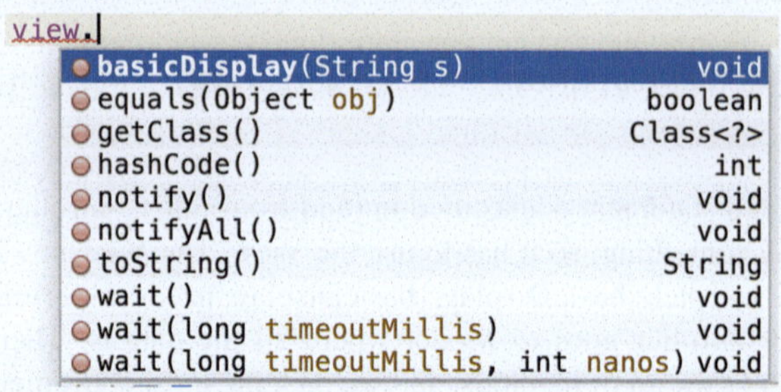

We picked the basicDisplay method which is a method that gives a very simple display of a string. This method requires a String as an input parameter.

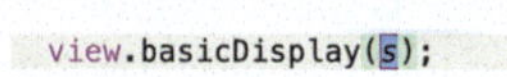

Then I tell the students we need some data, data represented as String, which is what is required by the basicDisplay method.

```
method basicDisplay in class View cannot be applied to given types;
  required: String
  found:    no arguments
  reason: actual and formal argument lists differ in length
----
(Alt-Enter shows hints)
view.basicDisplay(                );
```

To find some data, we need to go to the Model using the model object.

```
view.basicDisplay(    model.|            );
  getData(int n)                          String
  toString()                              String
  equals(Object obj)                      boolean
  getClass()                              Class<?>
  hashCode()                              int
  loadData()                              void
  notify()                                void
  notifyAll()                             void
  wait()                                  void
  wait(long timeoutMillis)                void
  wait(long timeoutMillis, int nanos)     void
```

From the available methods, we pick one that returns data from one specific data object, the one represented by the number 0.

```
view.basicDisplay(    model.getData(0)          );
```

This completed statement is executed in (and from) the Controller. A Controller needs to display something, so it has to use the View. The View has some methods available, and we pick basicDisplay. basicDisplay needs a String as an input parameter, so Controller goes to Model where all the data is. Then, we can see that Model offers a swath of methods, and we pick the most convenient for the task, getData. getData has different options, and we go with the number 0 which gets us the first available record from a list.

This statement encompasses all the aspects of MVC. The main role of Controller is to interact with Model and View. It highlights the role of View as displaying data. And also, the role of Model as the source of data. In summary, even in more advanced courses, if we can find the right basic concept, we can keep a minimalistic approach.

Reflections

The main challenge for an instructor is often resisting cramming too much content in a lesson or a semester. Another challenge is finding the balance between what you want to teach and what can be taught in a semester considering the program you are in, your students, and the available resources, such as access to computers and the availability of teaching assistants.

When initially designing this programming course, I started with problem-based learning (PBL). My end goal, which I never achieved, was to propose programming problems and let the students tell me what they needed to learn. I ended up with a more structured version of PBL. I spend half of the semester with weekly problems in a setup like the one described here. In the second half, I let the students build a project with minimum requisites specified.

The current version came from developments in the curriculum. Initially, this course was more advanced and represented the last major programming course in the program. In the redesigned curriculum, the course became an intermediate between a very basic Java course (no data structures such as classes, only basic Java statements) and a more advanced Java course.

I designed the course to focus on the most basic concepts that students needed to move to the advanced course. And for them to learn these concepts, I tried to remove everything else that would stand in their way.

What I Didn't Teach

In using a minimalist approach, I opted out of some topics. Although they are important, I saw them being in the way of an immediate grasp of basic concepts which was what I was aiming for.

For instance, some important but advanced uses of the IDE such as debugging or hitting a breakpoint were left out. Even formatting wasn't enforced although encouraged. These important topics were left for the next Java course. Also, the development of a cleaner code avoiding warnings from the IDE was not enforced although they were explained when students asked.

When my daughter was beginning to learn spelling, she came home one day with her notebook and had spelled favorite as *favrit*. Her grade schoolteacher had praised her spelling. I didn't know about phonics, and I couldn't understand what was happening. Later I was very surprised to see that, with time, she learned to spell and write correctly.

When people asked me how I could let students keep working with a few minor errors and avoid focusing on the details, opting instead to look at the big picture, I noticed the similarity between phonics and what I was doing.

The similarity is not in the method per se but in trying not to burden the new learner with unwanted details and rules. Instead, we can give young learners a way to grasp the most basic meanings of what they are trying to learn, providing a foundation for deeper understanding.

Reference

Fonseca, F., & Spence, L. (2014). The karate kid method of problem based learning. In J. Carroll (Ed.), *Innovative practices in teaching information sciences and technology*. Springer. https://doi.org/10.1007/978-3-319-03656-4_2

Teaching Information Technology Without the Technology: An Unplugged Approach to Introducing IT Concepts to First-Year Students

Chris Gamrat

Abstract This chapter discusses activities designed to introduce fundamental concepts of an introductory information technology course to first-year students or students taking the course as an elective. The activities are shaped by the Computer Science Unplugged area of research and pedagogy (Bell, Vahrenhold, CS unplugged—how is it used, and does it work? Adventures between lower bounds and higher altitudes: essays dedicated to Juraj Hromkovič on the occasion of his 60th birthday, 497–521, 2018). Across the chapter, activity topics include social network analysis and data visualization, cryptography, algorithms, and the binary number system. Each activity is explained, then the activity is connected to the curriculum of the course, and finally there are additional notes for how other instructors might use the activities in their classes. The chapter concludes by calling for further development and testing of this type of activity as a way of recruiting more diverse students into the field.

Keywords Computer science unplugged · Introductory IT · Low-tech · Applied concepts

Introduction

I developed, piloted, and taught using unplugged (i.e., paper-based) cyber activities in our introductory course, IST 110. Since 2018, I have used these materials in Penn State's College of Information Sciences and Technology (IST). While teaching introductory technology courses and engaging with prospective students, I have observed that the college regularly experiences the challenge of helping new and prospective students understand what IST is. I approached this need as an instructional design challenge to easily convey what sort of work is done in

C. Gamrat (✉)
Pennsylvania State University, University Park, PA, USA
e-mail: cwg118@psu.edu

J. M. Carroll (ed.), *Innovative Practices in Teaching Information Sciences and Technology*, https://doi.org/10.1007/978-3-031-61290-9_5

the college. This presented a unique opportunity to create materials for novices in the field that are low-cost materials, minimal technology, and a low barrier to entry to explore core concepts in our college. I first began teaching and encountered this need. My solution was to create paper-based activities that allowed an easy introduction that could be introduced early in the semester to give students exposure to ideas they would be covering in greater depth later in the semester. These materials support basic concepts within social network analysis, programming, cryptography, and the binary number system. The chapter details each activity, how it supports the students, and offers considerations for other instructors to use a similar technique.

Developing Low-Tech Solutions Informed by the Literature

These activities have been informed from a research area known as Computer Science (CS) Unplugged, a pedagogical approach to teaching computer science concepts traditionally for elementary and secondary students and in informal contexts (Bell & Vahrenhold, 2018). This technique is used as a mechanism for helping students see the fields of computer science and information technology as more approachable. However, there is also criticism in evaluating such pedagogical techniques. It also requires additional instruction to help students see how the concept applies to the field. Without that support explained by the instructor, students miss the connection due to limited prior knowledge.

Across the literature, there are many varied techniques for CS Unplugged, including several card games (Ford et al., 2017; Jin et al., 2018; Anvik et al., 2019) and a series of activities that formed as a practice "capture the flag" exercise to teach cybersecurity. I applied this concept to my introductory course to create class activities that help students explore concepts when they may also still be adding or dropping the course. While developing each, I wanted an activity highlighting concepts from across the college's curriculum.

Unplugged Activity Discussion

The following section details the activity, what it does for the students, and how other instructors can apply the ideas to their teaching.

Introducing Social Network Analysis and Data Visualization

The first activity offers students a way of approaching a dataset. Leveraging the analysis work of Gabasova (2016), this sheet shows students some key concepts

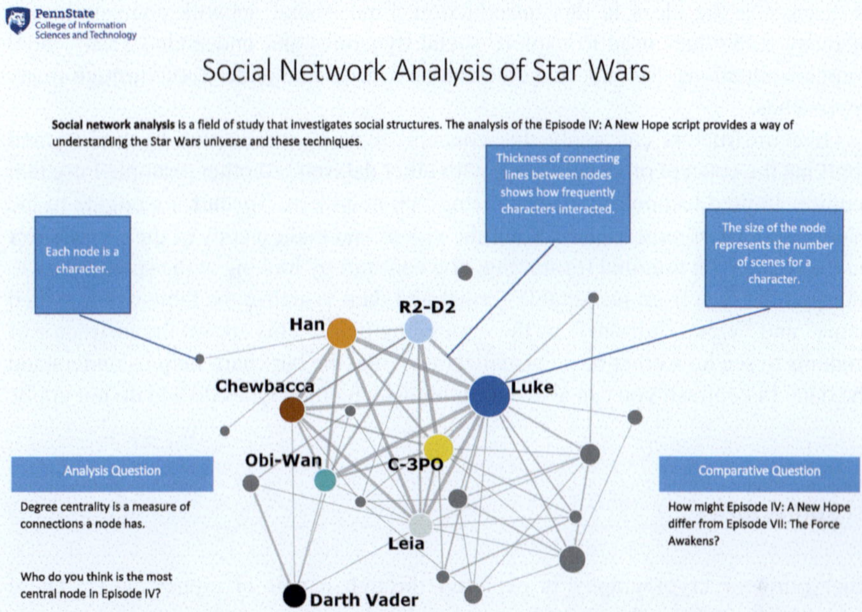

Fig. 1 Worksheet explaining social network analysis concepts utilizing an analysis of the interactions between characters in Star Wars: A New Hope

with a social network analysis and presents students with some initial questions for reflection. Figure 1 also explains to students that there are many ways of approaching data that are not what they might initially consider. This introduces concepts of general data visualization. When datasets are large, complex, or challenging to read, social network analysis can offer insight into the data.

Student Application of Social Network Analysis and Data Visualization To apply what was learned from reviewing the sheet, students are asked to discuss the prompts, such as how this would apply to other datasets such as other TVs, movies, and book series. Students can reflect on a visualization of character interaction in another example versus the visualization from Star Wars. Frequently, this prompts discussions around examples of other multipart books/movies such as Harry Potter. Later in the semester, we talk about big data and approaching datasets with a hypothesis for what may come from the data analysis and applying techniques that make sense for the dataset.

Considerations for Other Instructors While social network analysis is admittedly challenging, however, I would suggest something like this as it captures the concept with a relatively small number of individuals interacting. It is also important not to let the example of the movie overshadow the reason for the example. Using popular culture examples such as Star Wars (my students also like to use Harry Potter or the Marvel Cinematic Universe) is incidental. Rather, the essential element is discussing the ideas in this introduction. First, social network analysis is one of many techniques used to explore social dynamics and understand vastly more complex situations. Second, that data can be viewed and understood through many approaches.

Other instructors can apply this concept of data visualization by finding and applying the concept of visualization with other datasets. Another example I use that requires limited technology is visualizing "viral" videos. To start, I navigate to the Wikipedia List of most-viewed YouTube videos and make a copy of the spreadsheet on the page. With minimal formatting, you can start by looking at the spreadsheet to interpret the data from just a table format and then visualize the table using "video name" and "views (billions)" as the organizing fields. This allows the instructor or students to see how other basic visualizations, like the bar chart, help us understand the data. In contrast, you can also determine which graph types do and do not apply.

Cryptography

Encryption or cryptography is explored through a pair of cipher cards. These cards shown in Figs. 2 and 3 (designed to be about the size of a coaster) provide unplugged introductory information about types of ciphers. On one side of the card, an explanation is given for the type of cipher, and on the reverse side, there is a challenge to practice the cipher with an encoded message. The current design of the cards is intended to be a "call and response" or correspondence between the two. When decrypted, Fig. 2 asks the question, "Where shall we meet," and Fig. 3 responds with, "meet me at the creamery" (the Creamery is a popular location at the Penn State campus). Finally, the cards are designed with scaling in mind. A future

Fig. 2 Card front and back illustrating the concept of a Caesar cipher and the prompt to decipher a code

PennState
College of Information
Sciences and Technology

Rail fence cipher is a *permutation cipher* that follows that retains the original letters of the message but rearranges them. It staggers the message between two or more rows and then reads it off to give the encoded message.

PennState
College of Information
Sciences and Technology

Can you decode this two-row rail fence cipher?

M E M A T E R A E Y
E T E T H C E M R

Fig. 3 Card front and back illustrating the concept of a rail fence cipher and a prompt to decipher a code

iteration of the cipher cards would include more basic ciphers and demarcations for difficult levels.

Student Application of Cryptography This set of activities in the first week of class acts as a primer for several discussions around encryption throughout the term. For example, in my IST 110, students revisit the concept of encryption in lessons around computer networks, the Internet, and privacy. When we discuss these topics, we can refer to the basic unplugged activities to remind the students of how encryption fits within these topics.

Considerations for Other Instructors Using this unplugged activity can be developed by generalizing and through additional application/exploration of the techniques. First, the activity could be more generalized by not referring to an institutionally specific location. Given that this exercise references a Penn State location, it might not make sense to others. Students could explore this topic more by creating their own exchange of messages by using the ciphers in the document or by finding other similar techniques to expand their knowledgebase in the topic.

Algorithms

The activity in Fig. 4 walks students through the unplugged process of manipulating a dataset using the bubble sort technique written in plain language. The numbered squares at the bottom of Fig. 4 are intended to be torn off and used as cards to move around through the practice of running the bubble sort. This process allows students to see a slowed-down version of how the code would execute on their computers.

Student Application of Algorithms In IST 110, students explore concepts around application development, chiefly, the utility of pseudocode for outlining logic and

PennState
College of Information
Sciences and Technology

Sorting Techniques

Understanding, manipulating, and interpreting data are important skills developed in IST. This activity will provide an exercise to demonstrate how computers use different **algorithms** or techniques to sort data.

Bubble sort is a classic algorithm to sort small sets of data like this. Try to use bubble sort to get the cards organized. If there were more cards what might be a faster way to solve the puzzle?

Bubble Sort

1. To begin, tear off the cards, shuffle, and lay out the card in a line.
2. Compare the first two cards.
3. **If** they're not in numerical order, **then** swap.
4. Otherwise (**else**) move on to the next pair.
5. Apply step 1 to every following pair.
6. Work across the row completely and then return to the beginning.

1	2	3	4	5	6

Fig. 4 Sorting techniques worksheet walks students through conducting the manual process of executing the bubble sort algorithm

determining how the code will run. By discussing the application development process (and the many methodologies), students also understand how pseudocode fits into the development process as a time-saving exercise. Students can see that pseudocode, written in plain language, provides an opportunity to make their understanding of the code visible and testable before a single line of code is written.

Considerations for Other Instructors To further enhance this lesson, other instructors may also want to consider the following options in their own teaching through comparing algorithms and additional unplugged documentation.

There are numerous websites that offer animation examples of multiple sorting algorithms executing at once. These animations illustrate different approaches to the same problem, offering a way to think about both the complexity and efficiency in code as well as establishing that many algorithms for sorting are well-established and, once learned, can be applied in any number of future programming projects.

Instructors may also want to expand the lesson of application development through other forms of unplugged documentation. One such exercise that lends itself well to this sort of work is diagramming. Diagramming out the solution for the problem at hand can help students to understand the flow of information and various use cases for the application. Finally, instructors may also opt for expanding the lesson to also include a wireframe exercise to allow students to envision the interface of an application they are developing.

Binary

The concept of a binary number system is a foundational concept in understanding how computers work. I have introduced two activities to support learning the concept of binary for my formal teaching and student recruiting and outreach events. In the summer of 2023, these activities thoroughly tested in at the National Girl Scout Convention in Orlando, FL. During this event, my colleagues and I taught this less to over 2000 Girl Scouts, troop leaders, and parents.

Figure 5 illustrates an activity for students to encode their birthdays (a meaningful number) into binary. In the exercise, they do this for both the day and the month. For example, the student might identify that their birth month is September. By looking at the example reference sheet, they would know that September (the ninth month) can be converted into 10010 (we are using five bits for both the month and day because this is how many fingers most people have and because when totaled together across the five bits we have a sum of 31, the most number of days in a month). In the 10010 example, each value place is worth twice as much as the value before, so in total the nine for September is represented in the values of eight and one combined.

Figure 6 illustrates how students can count in binary in the same way we might learn to count in the decimal system: on our fingers! Each finger represents a bit of information and has either an off or on value. This is, when counting on your fingers, the finger is either up or down. I then explain that each bit is double the value of the

Fig. 5 Illustration of the process for identifying a student's birth date (month and day) and converting the information into binary represented by different colored beads (blue = one, white = zero)

CAN YOU COUNT TO 31 ON YOUR HAND?

It sounds impossible to count to 31 on one hand, but you can do it with binary code!

To count in binary code, each of your fingers gets a different value shown below. When a finger is up (or "on"), you count it. When a finger is down (or "off"), you don't count it.

- Thumb = 1
- Index finger = 2
- Middle finger = 4
- Ring finger = 8
- Pinky = 16

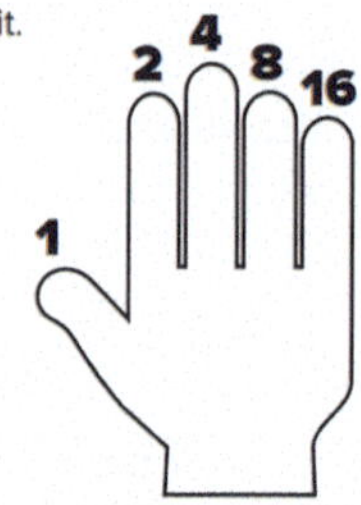

For example, if you have both your pinky and thumb up and your other three fingers down, that's 16 + 1, or 17.

If all of your fingers are up, you've counted to 31!

Fig. 6 Illustration of how to count using the binary number system on a student's fingers

prior bit since the value of all prior bits combined is *n*-1. This exercise works best as a follow along or with anticipatory questions such as "what comes after this?"

These activities work together in tandem to illustrate the concept of binary numbers.

Student Application of Binary Students apply the concept of binary when they learn about how computers work and how information is stored and transported. Across the term in IST 110, I revisit binary to describe how the binary number system is used. First, we discuss that as electronic devices, computers operate with the binary system represented through electrical "on" and "off" signals. Units of measure in this system are also revisited when we discuss storage (a hard drive having a capacity of gigabytes) or Internet speed (a modem supporting megabits per second). Once students understand what binary data represents, we can also engage in discussions about efficiency in web development and why you only want to include large files when you need to, why many modern websites don't fully load unless you continue to scroll through the content, and the benefits of streaming multimedia files.

Considerations for Other Instructors Other instructors may also want to consider including a discussion around data capture in their lessons. For example, I discuss the difference between digital storage and analog recording, explaining the difference between an MP3 file and a vinyl record. You can illustrate the digital capture of audio is a series of sampling at discrete points which is then converted into a value, whereas vinyl record produces a series of continuous grooves in the material. In prior instruction, this example also seemed to resonate with students.

Conclusion

These activities further the CS Unplugged curriculum development movement and are intended to make information technology topics more accessible to new students who may not have previously considered a technology major, thus widening the door to a broader, more diverse workforce technology-related fields.

References

Anvik, J., Cote, V., & Riehl, J. (2019, February). Program wars: A card game for learning programming and cybersecurity concepts. In *Proceedings of the 50th ACM technical symposium on computer science education*, pp. 393–399.

Bell, T., & Vahrenhold, J. (2018). CS unplugged—How is it used, and does it work? *Adventures between lower bounds and higher altitudes: essays dedicated to Juraj Hromkovič on the occasion of his 60th birthday*, pp. 497–521.

Ford, V., Siraj, A., Haynes, A., & Brown, E. (2017, March). Capture the flag unplugged: An offline cyber competition. In *Proceedings of the 2017 ACM SIGCSE technical symposium on computer science education*, pp. 225–230.

Gabasova, E. (2016, January 25). Star Wars social networks: The Force Awakens. *Evelina Gabasova Blog*. August 8, 2023, http://evelinag.com/blog/2016/01-25-social-network-force-awakens/

Jin, G., Tu, M., Kim, T. H., Heffron, J., & White, J. (2018, February). Game based cybersecurity training for high school students. In *Proceedings of the 49th ACM technical symposium on computer science education*, pp. 68–73.

Scenario-Based Design with ChatGPT: A First-Year Seminar Experience

John M. Carroll

Abstract A first-year seminar for university students has the goal of orienting new students to resources and opportunities of the campus. Ideally, students emerge from a first-year seminar feeling more comfortable, more situated, and more prepared to participate and succeed. In my 2023 first-year seminar classes, I incorporated a couple two approachable but significant content ideas from information sciences and technology, namely, the scenario-based design method and the ChatGPT artificial intelligence tool. I leveraged these ideas both in presenting various seminar topics and in shaping a mini-course project for the students. I hoped that this would help raise questions about concepts and methods throughout the seminar, thus making it a bit more like an academic seminar, and that the mini-course project would allow the students to experience group work, a commonly used teaching method in information sciences and technology.

Keywords ChatGPT · Problem-based learning · Scenario-based design · First year seminar

Introduction

In the fall of 2023, I taught three sections of the first-year seminar of Penn State's College of Information Sciences and Technology (IST). The concrete objective of a first-year seminar experience for university students is one of orientation. Students are not merely joining new classes and other activities, but they are joining an entirely different sort of learning community with hugely different resources and possibilities, as well as different practices and expectations. The seminars I taught met for 50 minutes per week throughout a 15-week semester.

J. M. Carroll (✉)
College of Information Sciences and Technology, Pennsylvania State University, University Park, Pennsylvania, PA, USA
e-mail: jmc56@psu.edu

J. M. Carroll (ed.), *Innovative Practices in Teaching Information Sciences and Technology*, https://doi.org/10.1007/978-3-031-61290-9_6

At the heart of the first-year seminar is a basic introduction to concepts, services, programs, and activities at the university. Class sessions address key orientational questions: What is academic integrity? What is academic advising? What are available majors, minors, and other academic programs and opportunities? How should a student think about and engage with various kinds of academic support, such as tutoring? How can a student make best use of the library system, information technology tools and resources, and other infrastructure at the university? What is research activity in the university, and how can undergraduates participate meaningfully in research? How can one improve one's learning strategies? How can students cope with, manage, and even leverage the stress they experience? What does it mean to have a proactive plan for self-engagement at the university? How does one plan for a semester or a year abroad? How does one start to prepare for a career?

Students can find relevant information on university websites, though, in my experience, far from a majority actually do that. But what makes a first-year seminar a seminar is that the topics are embedded in or at least related to the sort of everyday teaching and learning interactions that go on in seminar-type classes, namely, raising and addressing issues of concepts and methods with an instructor and with a relatively small group of fellow students. The seminars I taught in 2023 each had about 20 students, large enough to evoke a variety of views and sustain discussions, but small enough so that everyone had the chance to speak up. For the most part, each course meeting in my seminar was directed at one question or issue (as in the preceding paragraph). Usually, there was a short pre-class reading assignment and a post-class homework to strengthen and expand on what was discussed in class. About half the classes also involved guest speakers from a wide variety of campus organizations.

IST in Two Hours

The first class for the seminar focuses on brief presentation about IST and the first-year seminar in IST, the idea of a syllabus, a week-by-week course plan, and course policies and procedures regarding attendance, participation, grading, late work, academic integrity, accommodation for learning disabilities, and more. The course plan for the first-year seminar includes a core of about 6 weeks—academic integrity, academic advising, IST academic programs (2 weeks), internships and careers, library search skills, and diversity, equity, and inclusion (DEI). There is considerable flexibility about the remaining classes. My seminars included sessions on tutoring support, undergraduate research participation (2 weeks), student engagement, metacognition and academic success, and managing stress. Week 5 was a session on scenario-based design and my plan for the course project assignment (more about this presently); week 14 was a session in which student groups made an oral presentation of their course projects.

In the first meeting, I discussed what IST is about, to remind the students that the big picture of the class is one of orienting them to their new role as university students in the College of IST and to establish initial common ground for the course. I emphasized that IST embraces a broad view of what information technology is, including the sociotechnical perspective that technology is what people do with it and what they experience through interacting with it. I briefly introduced scenario-based design (SBD), which I defined as using narrative descriptions of human-technology interaction as core tools in planning for, envisioning, reasoning about, prototyping, and evaluating information artifacts. I shared simple but concrete examples of how SBD was used through the past 30 years to address difficult problems in information technology design and development, such as keeping the needs and concerns of users vividly in mind throughout the entire system development process. I also described my own role in developing SBD concepts and methods (e.g., Carroll, 1995) as a way of positioning myself personally in the course and in IST.

I also briefly introduced ChatGPT in the first class as a current awesome example of how new technologies and tools regularly upset the apple cart of standard IST problems and approaches. I mentioned that ChatGPT provides new ways to think about interactive applications and application design, that the course would include a group project in which students would design a novel ChatGPT application, and that details would come in later classes. This was partly to inspire the students and partly to get their attention. I did not want to worry them about the course project, at least not in the initial class meeting, but I wanted them to think of the seminar as seriously incorporating IST content.

I tried to employ SBD and ChatGPT in the class generally. For example, the second class (in week 2 of the semester) addressed academic integrity including professional ethics, the College Honor code, the variety of ethical misbehavior, and sanctions entailed by ethical misbehavior. Academic integrity is a key orientational topic for the first-year seminar, so I placed it up front. But I also wanted to help the students continue to think about SBD and ChatGPT. As a pre-class reading (that is to say, positioned in the week between the first and the second class), I shared a transcript of ChatGPT responding to this prompt:

Could scenario-based design help us analyze potential academic integrity issues that arise when students make use of ChatGPT?

ChatGPT made a series of seven suggestions, starting with the identification of typical ChatGPT interaction scenarios and analysis of how those scenarios contribute to academic activities, though in some cases undermining academic integrity (as when material is generated and copied, but not understood and elaborated). ChatGPT suggested that scenarios could be effective in helping students learn critical distinctions among different ways they can interact with ChatGPT. For example, idea brainstorming with ChatGPT can support a student's investigation, where generating and copying entire answers leaves nothing for the student to do. One is a legitimate use of an AI tool, but the other is cheating. ChatGPT also suggested that scenarios could be used to illustrate gray areas in current institutional policies to help initiate further discussion and policy development and so forth.

At the end of the second class, I shared a ChatGPT application concept with the class, as an example of the direction they could be thinking about for their course projects. I did this by prompting ChatGPT:

Imagine a smartphone app that integrates water quality monitoring data into Google Maps. Users could ask questions about surface water quality and risks at specific locations, including their current location. Is this something that ChatGPT could contribute to?

In response, ChatGPT identified six areas of design to which it could contribute, including scenario development. I then prompted ChatGPT with a scenario for the proposed app, which it elaborated. (The app concept for HydroGPT was developed collaboratively in a research proposal with Dr. Srishti Gupta of Columbia University.) I mentioned to the students that they could also work with ChatGPT to develop and refine their project design concept.

My argument is not that a couple hours on IST content in a 15-hour course is the right amount for a first-year seminar in IST. And it is important to emphasize that my students were all taking 4–5 other courses in the same semester, many of those courses providing insights into the foundations of IST. My argument is that even if most of the seminar is focused on important orientational concepts, services, programs, and activities at the university, some of it can be directed at key ideas from IST content. Moreover, this is a sound concept: it keeps the first-year seminar orientation tied to at least a couple content ideas from the students' academic program, reminding students and faculty throughout the semester that we are a group of IST scholars.

Scenario-Based Design

Scenario-based design was developed in the early 1990s. It was an early example of human-centered design, the idea that human characteristics and concerns can and must drive the envisionment of information technology, not merely the evaluation of technology intended for humans. SBD moves the focus from assessing and assuring the suitability for people of something already designed to conceiving and planning systems and applications that enable activity people can imagine and want to engage in Carroll (1995). It is a simple idea that leverages universal skills people already have to create, enjoy, and critically discuss narratives. This is incredibly powerful in that it opens the activity of designing information technology to all stakeholders, making it more possible for all stakeholders, technologists, and others to participate in the design process.

In the early 1990s, most system design work was predicated on the assumption that systematic descriptions of how people will interact with a system or a technology could effectively guide design, specifying interactions to be as simple and consistent as possible. Through the 1990s, such task analytic approaches were increasingly seen as tedious and inflexible but also as not sufficiently rich to address the design issues and objectives that mattered most to users (which is to say, real people). By the early 2000s, task analytic design was declining in favor of approaches, like scenario-based design, which emphasized flexibility, participation, and narratives of experience (Carroll, 2000, 2002; Hassenzahl, 2010; Wright & McCarthy, 2010).

I presented scenario-based design (and ChatGPT) to my class during the first class meeting, the one focused on explaining what the course would be about and how it would be conducted. These were brief introductions. For SBD, I emphasized that it illustrates how concepts emerge in IST to address key domain challenges, such as supporting design envisionment by bringing users, and their activity and experience to life in design discussions. I noted that Google Scholar identified fewer than 50 uses of the term "scenario-based design" before 1995, an average of 100 uses per year during the period 1995–1999, and an average of 400 uses per year since 1999.

I reviewed my 1999 paper "Five reasons for scenario-based design" describing how SBD allows designers to routinely go where design has never gone by imagining new approaches vividly and critically (Carroll, 1999). I used accessible examples, including virtual schools (Carroll et al., 2000), and the design of the Social Security Administration's website (usability.gov, 2024, no date) but also mentioned the famous "accidental war scenario" that shaped actual Cold War strategies (Carroll, 2016; Kahn, 1962).

In the fifth week of the course, I spent the entire class hour on SBD (and ChatGPT). As preparation, I assigned to read reflections that a group of Dutch app designers had posted online about their own adoption of SBD (Collard, 2014). My students carried out an online discussion based on that reading, identifying key strengths and weaknesses of SBD. I gave a lecture on "building the right system" versus "building the system right," expanding on the remarks I had made in the first class. Many IST undergraduate students see themselves as moving toward a professional career in which they design, analyze, and consult. Our students are often engaged by assignments that ask them to critically analyze concepts and methods in a realistic context of professional work.

In that class, I also presented another ChatGPT interaction, initiated with the following prompt:

Here is a design envisionment scenario for middle school science projects: "Marissa is a middle school science student. During a free period, she goes to the virtual physics lab, and finds two other students to discuss the role of the universal gravitational constant in planetary behavior. The students conduct several simulation experiments, and write a report." How can we take the next step in this design project to create a new computational infrastructure for middle school science?

I actually posed this prompt in two different ways to ChatGPT; it tends to produce long lists of answers, at least to my prompts. It identified a number of specific actions that could be taken to elaborate and further analyze the Marissa scenario. In my second prompt, I tried to get it to suggest what (single) first step I should take in developing my design. But it still gave me a list. I found this interesting as a concrete example of "prompt engineering" and the potential significance of slight changes in prompt language. It certainly showed the students that a brief scenario can evoke and guide a lot of design thinking. I again mentioned to the class that this example was quite similar to what they would be doing in their course projects. (This interaction with ChatGPT is the Appendix to this paper.)

At the end of the SBD class, I made an assignment for the students to develop a scenario to envision a future software system, application, or service—a brief narrative of around 100 words describing at least one human engaging in an activity supported by information technology. I asked them to present their scenario to ChatGPT with this prompt "How can this scenario help us identify specific technological requirements for developing this new information system?" Finally, I asked that they describe (in less than 100 words) what was learned about the scenario from ChatGPT's analysis.

This was the only class focused on explicit instruction regarding SBD and ChatGPT. These topics came up through the balance of the class, generally through student questions about the course project or through my making connections to other course topics. Indeed, the only other class meeting focused on SBD and ChatGPT was the 14th class in which the students presented their course projects.

ChatGPT

2023 was the year many people came to know a new kind of AI. ChatGPT 3 was launched in November 2022 as a free Internet tool by the research company OpenAI and quickly emerged as a popular embodiment of the threats and possibilities of AI (GPT3 is the third version of OpenAI's generative pretrained transformer, a language model that predicts the best response to language input (a prompt); https://openai.com/chatgpt). But ChatGPT could hold attention much better than a Terminator movie. Throughout the year 2023, new reports of its capabilities appeared in media, sometimes hysterical, sometimes inspiring. (In March of 2023, OpenAI released ChatGTP 4, a more advanced and licensed version; all references in this essay are to ChatGPT 3 and 3.5, the original and freely accessible version, first released in November 2022.)

Its actual capabilities expanded week by week as well. For example, early in the spring of 2023, ChatGPT was drawing fire from critics for projecting too much self-confidence in its suggestions. It deserved the criticism, people were prompting it with complex challenges, and many of its responses were lame. Only weeks later, users noticed that it had begun to adopt a more modest voice and to qualify its responses. For example, in giving interactive system design advice, it appended a reminder that if one wants to design an interactive system, one must necessarily involve domain experts and get feedback from representative users. These enhancements are sensible and typical of advice-giving strategies used by humans (Carroll & McKendree, 1987).

ChatGPT is unlike AI capabilities that people were most regularly encountering in 2023, such as recommendation algorithms and conversational assistants. Recommendation interactions are implicit; AI software identifies actions and patterns in a person's behavior that indicate interests or preferences. For example, years ago, I discovered a tight-fitting shirt for dogs that snuggly reassures frightened or upset canines, allowing them to stop barking or bark less. I think it was called a Thunder Shirt. I investigated it once, decided it was not the solution I needed, and looked into it no further. But that one episode impressed a recommendation agent, and for months, I noticed Thunder Shirt ads and other connections popping up in my browser. Conversational agents, like Siri and Alexa, offer direct natural language interactions, but they are extremely limited in what they can handle. They can be useful if a person has simple needs like timing an egg or hearing Bob Dylan's Christmas album, but for most topics, they tend to reply "I don't know that."

As an interactive system experience, ChatGPT presents itself modestly, as a teletype. To me, it nostalgically evoked Weizenbaum's script-based ELIZA chatbot from the 1960s, a classic reductio ad absurdum of AI chutzpah (Weizenbaum, 1976). People using ChatGPT and ELIZA are initially presented with a blinking text cursor and were expected to type a textual "prompt." In Weizenbaum's system, users were limited to prompts and responses modeled on Rogerian psychotherapy interactions. Weizenbaum (1976: 3-4) published an early (and now very famous) example:

```
User: Men are all alike.
ELIZA: IN WHAT WAY?
User: They're always bugging us about something or other.
ELIZA: CAN YOU THINK OF A SPECIFIC EXAMPLE?
User: Well, my boyfriend made me come here.
ELIZA: YOUR BOYFRIEND MADE YOU COME HERE
User: He says I'm depressed much of the time.
ELIZA: I'M SORRY TO HEAR THAT YOU ARE DEPRESSED.
User: It's true I am unhappy.
ELIZA: DO YOU THINK COMING HERE WILL HELP YOU NOT TO BE UNHAPPY?
User: I need some help, that much seems certain.
```

In the mid-1960s, many people who engaged with ELIZA through a teletype user interface were convinced that a human was on the other end of the conversation. In the 1960s, no surprise, this suggested that ELIZA had passed the Turing Test of intelligence by being indistinguishable from a human interlocutor (Turing, 1950).

Weizenbaum himself was shocked by this reaction. Looking critically at the ELIZA interaction suggests what is going on "inside" ELIZA: each response is scripted to keep the conversation going without adding any content. Indeed, the first two responses would work about as well no matter what the user typed. The third response slightly transforms the user prompt, a rhetorical signal that the user should elaborate further. This evokes a prompt including for the first time in the conversation the trigger word "depressed." ELIZA has a standard empathetic response, which subsequently evokes another trigger word "unhappy."

ELIZA was not artificial intelligence in any significant sense. It was instead a clever demonstration of how a successfully interactive chatbot could rely on nearly no knowledge of language or understanding of the conversations in which it appeared to participate. In this respect, it echoed a long line of research in social psychology showing that people tend to conflate intentionality with animation

(Scholl & Tremoulet, 2000). But it was a major contribution to artificial intelligence research. It elegantly disposed of the Turing Test and reminded humans that they must consider more than utterly superficial indicators when they make the attribution that something is "intelligent."

ChatGPT turned debates about the Turing Test upside down. It did this by correctly answering simple fact-based quiz and homework questions, a staple of traditional education. I remember a sad plea on our faculty list serve for help from a young colleague at Penn State who had given her class an online quiz and was stunned that her students all scored A's and with very similar responses. Instead of pretending to perform like humans by imitating the veneer of intelligence, ChatGPT had exposed an actual weakness in human intelligence, in this case in pedagogy. In a world with highly capable information tools, asking students to merely retrieve information in quizzes and assignments is no longer appropriate. Throughout the Spring semester of 2023, questions and concerns were raised by university administration, by faculty, and even by students about challenges for teaching and learning and for academic integrity. Would anyone ever again write a paper or answer a quiz item entirely on their own, when a short conversation with ChatGPT could help so much?

Results

As described above, three student assignments in my first-year seminar were specifically directed at SBD and ChatGPT. I looked at what students produced in these assignments for indications that my inclusion of the two IST "content" topics engaged the students intellectually and supported learning. As preparation for the week 5 class on SBD, I assigned my students to contribute to an online discussion forum identifying one strength and one weakness of scenario-based design as a method for identifying and developing design concepts for new systems, applications, and services. They were to base their posts on the reflections of a Dutch design group that had adopted SBD but also on class discussions, their own ideas, and anything else they might find on the Internet by Googling "scenario-based design examples."

Most of the strengths the students identified came from the Dutch design team reflections, including that scenarios could help visualize a system before it is created, could make clearer to the design team what the system will be and do, and how that will be of value to users, and could help designers coordinate actions they take with the ideas and goals they are pursuing. Students of course had many ways of putting these points, such as "put yourself in users' shoes," "see their products from the user's perspective," and "keep team focused on where they want to end up." Students tended to focus on strengths that bear directly on designers and users, as opposed to bearing on design problems per se. For example, it is often noted that scenarios help focus design problems, issues, and conflicts by concretizing them and that they encourage iteration and alternative solutions by being malleable and inexpensive to change.

The weaknesses were less emphasized in the Dutch design team's reflections; they did make the point that adopting SBD required them to exercise new skills and that in a few instances they realized that they initially did more work than was required to get the benefits of SBD and that subsequently they allocated effort better. In general, they conveyed that SBD was an efficient method that improved their work output.

My students suggested that one cannot identify every important scenario and might identify incorrect scenarios. They noted that various stakeholders may have different opinions, and it could be difficult or impossible to decide who is right. They also suggested that it could be time-consuming to create user scenarios and consider user issues, that generating so much information might make it more difficult and slower to make decisions, and that focusing too much on thinking about design would undermine building something. The students' weaknesses tended focus on project management issues that could be entrained by lack of a general stopping rule for generating scenarios, by not having objective criteria for evaluating scenarios, and by the possibility that scenario analysis could slow progress toward implementation. I like the points they came up with, which were for the most part not ones I or the Dutch designers primed but which do reflect real issues in design and development work.

The second assignment I made that was specifically directed at SBD and ChatGPT was the homework from week 5. I had asked students to develop a scenario of about 100 words envisioning a future software system, application, or service and to ask ChatGPT how the scenario could help to identify specific technological requirements for developing an information system. The students came up with relatively interesting scenarios involving interactions with novel applications and services, for example, a music streaming app with social features that can make listening to music a shared experience; a thermostat reads the body temperatures of nearby people and dynamically heats/cools the room for comfort; a fitness app that is oriented to local communities and keeps track of what everyone does, maintaining leaderboards and reports for all kinds of activity to keep the community moving; a school sign in system for students and faculty that efficiently manages who enters the school and subsequently where each authorized person is; and a crowd monitor for public venues that manages load at various entrances in order to minimize queuing and wait times.

The students were enthusiastic about their experience reviewing scenarios with ChatGPT. Most of the students commented that it was helpful to see how their scenario envisionment entailed specific technology capabilities and policies that would need to be created or adapted in system development. However, although the students seemed to feel this feedback was guidance, in many cases students just listed or summarized the points ChatGPT made. (In a couple cases, they merely cut and pasted ChatGPT's output into their own paper, a problematic interaction ChatGPT itself had critiqued for them just a few weeks before in the academic integrity class.)

Students particularly liked specific suggestions ChatGPT made to enhance their scenarios: “It also brought up some good ideas that I would not have thought of to make the technology better for users such as saving personal preferences on temperature and possibly connecting this system to other smart home devices.” ChatGPT also helped “filter” novel design proposals that were in fact not so novel. One student had suggested a speech user interface scenario, but ChatGPT told him that this capability was already available and gave him a step-by-step procedure to set it up. He was disappointed but also impressed and intrigued. Other students suggested a thought-to-text transcription system and a personal distributed AI agent, respectively; ChatGPT delved into technologies that might be promising for developing these scenarios but also emphasized open sociotechnical issues that need to be confronted.

The third assignment I made that was specifically directed at SBD and ChatGPT was the course project, presented orally in class in week 14 and submitted as a final paper a week later. Projects involved working in groups of typically four students to develop 2–3 key scenarios. I emphasized that the scenarios should animate, motivate, and explain how and why people would interact with the envisioned app. Some of the week 5 scenarios had been a bit austere with respect to depicting people. I asked the groups to develop their designs by analyzing strengths and weaknesses: potential risks and benefits for users, for other stakeholders, and for society in general and other relevant design trade-offs. I suggested that they again work with ChatGPT to identify and think critically about issues and possibilities. I asked them to identify and analyze at least three design issues. I also asked them to make use of at least three general literature references, such as web pages and other mass media. (This was to revisit and strengthen skills from the week 12 class on library search skills.) Finally, I asked the groups for a brief description of who did what among members of the group, when in the process members made their contributions to the group work, and what notable problems or insights emerged from the group collaboration. (This was their first group project as university students.)

I emphasized to the students that the oral presentations were primarily for them to share their projects with and get feedback from myself and the rest of the class. The final written project papers were due a week later. Especially given that I emphasized their own goals were the primary goals for the presentation, I was pleased with the engagement and preparation that was displayed. For all of the groups but one (14/15), all of the members showed up and participated actively and knowledgeably in the presentations. The students clearly presented their design rationale and scenarios, identified strengths and weaknesses of their designs, engaged literature in describing context for their work, and referred to interactions with ChatGPT. About half of the presentations were notably energetic and even theatrical.

Some of the projects were further developments of the individual project concepts from week 5. For example, the thermostat that senses body temperature now incorporated ChatGPT’s idea about saving preferences, literature-based argumentation about how ambient temperature can support circadian rhythms, and a scenario with two people, each having different temperature needs and preferences. Most projects were new concepts, such as a universal tutoring service, a photo-based tourist information service that allows future visitors to build advance familiarity with places, and an AI coach that helps people organize their feelings and experiences into songs. A few concepts were enhancements of existing services such as roommate matching and ride sharing but with heightened expectations derived from potential AI capabilities. Several projects addressed student planning and study support issues, such as an assistant that gathers course artifacts, including syllabi, and constructs study guides, flashcards, and practice tests, iteratively refining its support to address any detected weak area in student.

I was impressed with the weaknesses/downsides students identified. Especially relative to the individual papers they had done for the week 5 assignment, they considered a far wider range of risks, such as system concepts that depend on sensing and recognition technologies which themselves might not perform well enough, personal liabilities potentially entrained by helping people arrange possibly dangerous activities like rideshares, incompatibilities among the many different security protocols now employed by American universities, challenges of attaining critical mass among users for designs depending on networking dynamics and of creating voluminous amounts of content for designs depending of rich information, as well as the downside of overreliance on AI services that could inadvertently undermine the competence of users, purvey bias and unfairness, or employ falsehoods. Several project teams mentioned that ChatGPT particularly helped them identify potential downsides of their scenarios.

I mentioned earlier that I, and many of my faculty colleagues, have found realistic group projects, like the one employed in these first-year seminars, are very effective at motivating students to inhabit the "IT professional" role that they envision for themselves after college. They seem more engaged and autonomous, and they take more responsibility. What is especially novel in this case is that the students were doing this right at the starting gate of their university studies, with relatively little scaffolding or handholding from me. I tried to consistently frame the projects in these seminars as "lightweight explorations" intended to expose the students to pedagogical activities they would encounter again throughout their trajectories as IST undergraduates. This framing was consistent with the tone of the entire first-year seminar, which beyond the course project consisted of relatively light modules addressing a potpourri of topics. Nevertheless, I found it quite encouraging and affirming of what I tried to do here. I don't see any reason not to engage IST students in realistic group-based projects from day 1 onward.

Indeed, pondering this course experience leads me to suggest that this model could be usefully broadened and extended to other first-year courses. The first year is often a time to take lower-level prerequisite courses, in science and mathematics, and general education courses in writing, communication, and humanities. These are important courses but the fact that they are typically configured to have no concrete relevance to IST topics cannot possibly make them more relevant to IST students. Instead, they are relevant and important to students despite their lack of concrete connection to IST. On the other hand, strongly specializing such foundational courses to IST (and other teaching areas) would not support flexibility; for example, many first-year students change their majors.

My thought is that first-year courses could include limited specialization that helps students more concretely understand and appreciate the contribution that foundational courses make to their education. My first-year seminars allocated about 3 hours in total to scenario-based design and ChatGPT (the third hour being oral project reports). I felt this modest amount of specialization helped me to connect general orientational topics more concretely to the IST content area and, I hope, helped the students both progress concretely as IST students and also to make more sense of the orientational content of their first-year seminar. It would be interesting to see how a model of "lightweight exploration" might help first-year students better understand and appreciate foundational courses more generally.

Acknowledgments I am grateful to Professor Jessica Kropczynsky who mentioned to me that ChatGPT "likes" scenario-based design. What this might mean initially puzzled me but led directly to the project assignment described in this chapter. I also thank Dr. Srishti Gupta with whom I developed the HydroGPT app concept.

Appendix: Scenario Development with ChatGPT

Here is an interaction I had with ChatGPT probing "scenario development." I asked about next steps after one has identified a key interaction scenario. ChatGPT gave me a list of things to do. So, I then asked what would be the first thing to do in developing a design? ChatGPT again gave me a list.

The answers are quite reasonable and should at the least help you think about questions and answers about scenario development in human-centered design.

Design Envisionment and Analysis

JMC

Here is a design envisionment scenario for middle school science projects: "Marissa is a middle school science student. During a free period, she goes to the virtual physics lab and finds two other students to discuss the role of the universal gravitational constant in planetary behavior. The students conduct several simulation experiments and write a report." How can we take the next step in this design project to create a new computational infrastructure for middle school science?

ChatGPT

Creating a new computational infrastructure for middle school science projects like the one you described is an ambitious and valuable endeavor. It involves combining educational technology, curriculum design, and a user-friendly platform. Here are steps you can take to move forward with this design project:

1. Needs Assessment and Research
    - Conduct extensive research to understand the specific needs and challenges faced by middle school science educators and students in your target audience.
    - Identify the curriculum standards and learning objectives that the computational infrastructure should align with.
2. Engage with Educators and Students
    - Collaborate with middle school science teachers to gain insights into their teaching methods and how technology can enhance the learning experience.
    - Involve students in the design process through surveys, focus groups, and usability testing to understand their preferences and needs.
3. Define Learning Goals
    - Clearly define the learning goals and objectives that the computational infrastructure should support. For example, in your scenario, it's understanding the role of the universal gravitational constant in planetary behavior.
4. Content Development
    - Develop interactive, engaging, and age-appropriate content that aligns with the curriculum standards. This content can include simulations, virtual labs, videos, quizzes, and other educational materials.
5. User-Friendly Interface
    - Design a user-friendly interface that is intuitive for both students and teachers. Consider the use of graphics, animations, and a clean layout.
    - Ensure the platform is accessible to students with various levels of technological proficiency.
6. Simulation Tools
    - Develop or integrate simulation tools that allow students to conduct experiments and explore scientific concepts. These tools should be accurate and provide immediate feedback.
    - Include a wide range of simulations covering various science topics.

7. Collaboration and Communication
    - Build features that enable collaboration among students, similar to your scenario where Marissa interacts with peers. This can include discussion forums, chat, and video conferencing.
8. Assessment and Progress Tracking
    - Implement assessment tools for teachers to evaluate student progress and understanding.
    - Create a system for tracking and reporting on student performance and engagement.
9. Teacher Support
    - Provide resources and support for teachers, including lesson plans, guides, and professional development opportunities.
10. Feedback and Iteration
    - Continuously gather feedback from teachers and students to improve the platform.
    - Regularly update the content and features to keep the platform relevant and engaging.
11. Privacy and Security
    - Ensure that the platform complies with privacy laws and regulations, especially when dealing with student data.
    - Implement robust security measures to protect user information.
12. Pilot Testing
    - Pilot the computational infrastructure in a few middle schools to gather real-world feedback and make necessary adjustments.
13. Scale and Deployment
    - Once the infrastructure is refined and tested, deploy it to a broader audience of middle schools.
14. Training and Support
    - Offer training and ongoing support to educators and students to ensure they can effectively use the computational infrastructure.
15. Sustainability Plan
    - Develop a plan for the long-term sustainability of the project, including funding, updates, and maintenance.

Creating a new computational infrastructure for middle school science projects is a significant undertaking, but it can have a profound impact on science education. Collaboration with educators, students, and experts in educational technology will be crucial throughout the development process to ensure the infrastructure meets the needs of its users.

First Steps in Developing the Design

JMC

What would be the first thing to do in developing this design?

ChatGPT

ChatGPT The first step in developing the design for a new computational infrastructure for middle school science projects is to conduct a thorough needs assessment and research phase. …

1. Identify Stakeholders: Determine who the key stakeholders are in middle school science education. This includes teachers, students, curriculum developers, and administrators.
2. Conduct Surveys and Interviews: Reach out to middle school science teachers and students to gather their input. Conduct surveys and interviews to understand their current challenges, pain points, and what they believe would enhance their teaching and learning experiences.
3. Review Curriculum Standards: Familiarize yourself with the curriculum standards and learning objectives that apply to middle school science education. These standards will help you define the educational goals your computational infrastructure should support.
4. Explore Existing Solutions: Research existing educational technology tools and platforms that are currently used in middle school science classrooms. Identify gaps and areas where your project can offer unique value.
5. Gather Feedback: Attend science education conferences, workshops, and meetings to engage with educators and gain insights from experts in the field. Their feedback can be invaluable in shaping your project's direction.
6. Consider Technology Accessibility: Assess the technology infrastructure in middle schools, including the availability of devices and internet connectivity. Ensure that your computational infrastructure is accessible to schools with varying levels of technological resources.
7. Identify Content Needs: Determine the specific science topics and concepts that are most challenging for students and teachers. Identify where interactive simulations, virtual labs, or other computational tools could provide the most value.
8. Legal and Ethical Considerations: Familiarize yourself with the legal and ethical considerations of working with student data and educational technology.

Ensure that your project complies with relevant privacy and security regulations.
9. Budget and Resources: Assess the budget and resources available for your project. Determine if you have the financial and human resources needed to develop and maintain the computational infrastructure.
10. Project Scope: Clearly define the scope of your project based on your research findings. Determine the specific features and functionalities you intend to include in the computational infrastructure.

By conducting this comprehensive needs assessment and research, you'll have a solid foundation for designing a computational infrastructure that addresses the genuine needs and challenges of middle school science education. It also helps ensure that your project aligns with curriculum standards and has a higher likelihood of being well received by educators and students.

References

Carroll, J. M. (Ed.) (1995). *Scenario-based design: Envisioning work and technology in system development*. Wiley.

Carroll, J. M. (1999). Five reasons for scenario-based design. In R. H. Sprague (Ed.), *Proceedings of the 32nd Annual Hawaii International Conference on System Sciences*, Maui, January 5–8. IEEE Computer Society.

Carroll, J. M. (2000). *Making use: Scenario-based design of human-computer interactions*. MIT Press.

Carroll, J. M. (2002). Making use is more than a matter of task analysis. *Interacting with Computers, 14*(5), 629–637.

Carroll, J. M. (2016). Theorizing the unprecedented. In D. H. Sonnenwald (Ed.), *Theory development in information: Reflecting on the process*. University of Texas Press.

Carroll, J. M., Chin, G., Rosson, M. B., & Neale, D. C. (2000). The development of cooperation: Five years of participatory design in the virtual school. In *Proceedings of the 3rd Conference on Designing Interactive Systems: Processes, Practices, Methods, and Techniques (DIS '00)* (pp. 239–251). Association for Computing Machinery. https://doi.org/10.1145/347642.347731

Carroll, J. M., & McKendree, J. (1987). Interface design issues for advice-giving expert systems. *Communications of the Association for Computing Machinery, 30*(1), 14–31.

Collard, M. (2014). *Scenario-based Design*. [Online]. Available: https://unitid.nl/2014/02/scenario-based-design/

Hassenzahl, M. (2010). Experience design: Technology for all the right reasons. *Synthesis lectures on human-centered informatics*. Springer Nature.

Kahn, H. (1962). *Thinking about the Unthinkable*. Horizon Press.

Scholl, B. J., & Tremoulet, P. D. (2000). Perceptual causality and animacy. *Trends in Cognitive Sciences 4*(8), 299–309.

Turing, A. M. (1950). Computing machinery and intelligence. *Mind, 59*(236), 433–460.

usability.gov (2024). [Online]. Available: https://www.usability.gov/how-to-and-tools/methods/scenarios.html

Weizenbaum, J. (1976). *Computer power and human reason: From judgment to calculation*. W. H. Freeman & Co.

Wright, P., & McCarthy, J. (2010). Experience-centered design: Designers, users, and communities in dialogue. In *Synthesis lectures on human-centered informatics* (Vol. 3, No. 1, pp. 1–23). Springer Nature.

Part II
Meeting Students Where They Are: Emerging Challenges

Hitting the Sweet Spot: Designing an Accessible Introductory Machine Learning Course for Informatics Students

Amulya Yadav and Hangzhi Guo

Abstract Machine learning applications are ubiquitous in today's postmodern societies, with applications in healthcare, criminal justice, urban planning, environmental sustainability, etc. This rising interest in machine learning (ML) is matched by an ever-increasing demand among college students to learn ML. Unfortunately, the most basic and commonly used treatment of ML in computer science curriculums around the country is quite technical and math-heavy, which demotivates students from nontechnical backgrounds to take up ML courses (despite their interest in learning how to use ML). Students in iSchools (or information schools) are characterized by their diversity of educational backgrounds and a great commitment to interdisciplinarity. As such, an ML course offering inside an iSchool cannot assume prerequisite mathematical ability among all students who wish to take this course. This paper discusses an attempt at designing an easily accessible ML course that was specifically designed for Informatics students at the Pennsylvania State University. In particular, this paper discusses an important trade-off between a math-heavy treatment of the course and a more use-oriented treatment of the course. Further, this paper sheds light on the design choices made during the curriculum design, evaluation rubric design, and content delivery of the course in order to help balance this trade-off and, in turn, cater to the diverse backgrounds and interests of Informatics students.

Keywords Machine learning pedagogy · Informatics · Applied ML

Introduction

Machine learning is a subfield of artificial intelligence and computer science which focuses on the use of data and algorithms to imitate the way that humans learn, gradually improving its accuracy. Over the last two decades, the field of machine

A. Yadav (✉) · H. Guo
Pennsylvania State University, University Park, PA, USA
e-mail: amulya@psu.edu

J. M. Carroll (ed.), *Innovative Practices in Teaching Information Sciences and Technology*, https://doi.org/10.1007/978-3-031-61290-9_7

learning (ML) has pervaded almost every aspect of day-to-day life in postmodern societies. For example, recent advances in ML have powered many applications of daily use, such as route navigation (via Google Maps), recommender systems (on Amazon and Netflix), and smart home assistants (e.g., Amazon Alexa, Apple Siri). As such, it is no surprise that the "global machine learning market has grown steadily: in 2021, it was valued at $15.44 billion, and owing to the increasing adoption of technological advancements, it is expected to grow from $21.17 billion in 2022 to $209.91 billion by 2029, at a CAGR of 38.8%" (Size, 2021). While the ML industry has made inroads into most world economies, "North America has accounted for 43% of the global AI market in 2022 due to high demand for automated and technologically advanced solutions and favorable government policies encouraging AI adoption" (Precedence Research, 2023).

This rapid growth of the global machine learning market has led to a concomitant rise in the demand for machine learning knowledge and skillsets among current and new members of the tech workforce. To match this ever-increasing demand for ML skillsets, many higher education institutions (such as universities and colleges) have created specialized degree programs on artificial intelligence, machine learning, and/or data science that aim to create a new well-equipped workforce that can respond to the rapidly evolving needs of the AI/ML industry. For example, the Pennsylvania State University offers an MPS (Master of Professional Studies) program in artificial intelligence (Penn State University, n.d.-a), while the College of Information Sciences and Technology (within the Pennsylvania State University) offers a data science major for undergraduate students (Penn State University, n.d.-b). Other universities which offer specialized AI/ML degree programs include Carnegie Mellon University (n.d.), Georgia Tech (n.d.), University of Michigan (n.d.), etc., and such programs are only expected to increase in number in the near future.

Unfortunately, designing curricula for such specialized degree programs in AI/ML is quite challenging because of two reasons. First, current curricula are mainly designed on the principle of "reuse," since ML and data science is traditionally housed within computer science and statistics departments; the prototypical curriculum of an AI/ML or data science degree program expects students to take some courses on AI/ML that are offered by computer science departments and some courses on probability theory, linear algebra, etc. that are offered by statistics departments. While certainly there are a lot of benefits with reusing well-designed course curricula from other allied departments, this design choice of "reuse" significantly hinders accessibility of the course materials by nontechnical students (I cover this point in greater detail in the paragraph below).

Second, the universal applicability of ML as a science has led to several applications and tools being spawned in domains as diverse as healthcare, criminal justice, city planning, agriculture, etc. As a result, the field of ML and data science has grown larger than its parent field of computer science (CS). As such, courses on ML and data science offered in universities draw interest from students across a wide variety of majors including computer science, human-centered design, electrical engineering, political science, and sociology (to name a few). Given the widespread disparity in mathematical and programming knowledge among students belonging to different majors and disciplines, it becomes a challenging proposition for any instructor to create a course offering that can keep students of all disciplines and backgrounds interested.

While a mathematical treatment of ML and data science topics may be interesting to students from CS majors, it may prove too daunting for students in human-centered design. Similarly, a nonmathematical treatment of ML may improve accessibility of the course to a wider audience; however, it may restrict the depth in which topics can be covered, thereby making the course somewhat superficial. This point about superficiality is particularly relevant in today's day and age, with several ML and data science offerings on popular massive open online courses (MOOC) platforms (like Coursera (n.d.), Udacity (2022), to name a few) specializing in giving a crash course on how to use ML, even though they might not help students in understanding why the ML models work and when they can be expected to work. This has led to an added pressure among university instructors to create course offerings on ML and data science (with technical depth) which provide added value on top of popular MOOC offerings.

This paper discusses an attempt at designing an easily accessible ML course that was specifically designed for Informatics students at the Pennsylvania State University. Students in the Informatics program at Pennsylvania State University are characterized by their diversity of educational backgrounds and a great commitment to interdisciplinarity. As such, an ML course offering inside an Informatics major cannot assume prerequisite mathematical ability among all students who wish to take this course. In particular, this paper discusses an important trade-off between a math-heavy treatment of the course and a more use-oriented treatment of the course. Further, this paper sheds light on the design choices made during the curriculum design, evaluation rubric design, and content delivery of the course in order to help balance this trade-off and, in turn, cater to the diverse backgrounds and interests of Informatics students.

Designing an Undergraduate Course Offering on Machine Learning for Data Analytics

Now, I will discuss the author's attempt at designing an easily accessible ML course that was specifically designed for Informatics students at the Pennsylvania State University. We structure this discussion into two stages. First, I provide background context about the Informatics major offered by the College of Information Sciences and Technology at the Pennsylvania State University, as this background context played an important role in my design choices for this course. Specifically, I will discuss how students in the Informatics program are characterized by their diversity of educational backgrounds, and a great commitment to interdisciplinarity, and how this meant that standard course offerings on ML would be insufficient for students in the Informatics program. Second, I will delve into my specific course structure and evaluation rubric. Finally, I will end by introspecting on lessons learned through the process of teaching this course and my suggestions at improving this course for the future.

The Informatics Major at Pennsylvania State University

The Informatics degree program (as referred to this in paper) broadly refers to all degree programs that are offered by the **Penn State College of Information Sciences and Technology** (also called the College of IST). In brief, the College of IST offers six bachelors of science degrees: (i) Cybersecurity Analytics and Operations; (ii) Applied Data Sciences; (iii) Enterprise Technology Integration; (iv) Human-Centered Design and Development; (v) Information Sciences and Technology; and (vi) Security and Risk Analysis. In addition, the College of IST offers three graduate programs: (i) the M.S. program in Cybersecurity Analytics and Operations; (ii) the M.S. program in Informatics; and (iii) the Ph.D. program in Informatics (Penn State College of Information Sciences and Technology, 2023). We refer to all these programs informally as "Informatics" degree programs in the rest of this paper, simply because they are offered by the Information School (or iSchool) at the Pennsylvania State University (i.e., the College of IST).

The common theme across all these degree programs offered by the college is that, broadly, they focus on the study of issues that exist "at the intersection of information, technology, and society, aiming to prepare leading professionals and scholars who can leverage technology and critical thinking skills to solve the complex challenges of an information society." These programs are designed to extend beyond "classic computer science, management information systems, and library science to prepare students to meet challenges in the use of computers and networked systems for applications such as intelligent systems, medicine, business, homeland security, environmental monitoring, and control of complex systems." The hope is that these programs can "produce graduates who would have basic knowledge of information technologies such as artificial intelligence, computer programming, discrete mathematics, database concepts, human-computer interaction, simulation, and understanding of information system concepts, as well as the capability to work in teams to understand how information technologies can be utilized in real applications involving individuals, organizations, and ultimately national or global enterprises" (Penn State College of Information Sciences and Technology, 2023).

Currently, work in the college of IST currently focuses on four major areas in education and research: (i) data sciences and artificial intelligence, (ii) human-computer interaction, (iii) privacy and security, and (iv) social and organizational informatics. As such, students are expected to choose a concentration out of these four focus areas when they enroll into a degree program. It should be noted that asking students to choose a concentration is not intended to be restrictive;

the goal is to give students a general direction for their studies, while affording them the flexibility of choosing courses from other concentrations as well. As a result, it is very common for a student majoring in Applied Data Sciences to take several courses from other concentrations, such as Human-Centered Design and Development, Cybersecurity Analytics and Operations, etc. Similarly, it is very common for students majoring in Human-Centered Design and Development to take courses from the Applied Data Sciences Curriculum.

The Challenge of Creating an Accessible ML Course Offering in the College of IST

The intention behind creating such an interdisciplinary degree curriculum is to hopefully create graduates whose knowledge and skillsets cut across a wide variety of disciplines, and thus, it would enable them to interface with a wide variety of experts in the field such as AI researchers, HCI experts, cybersecurity practitioners, etc. While this interdisciplinary nature of the Informatics degree program is undoubtedly a huge strength of the program (and the university), this emphasis on interdisciplinarity and cross-cutting education streams also creates hurdles and challenges for instructors who need to carefully design their course offerings such that all students at the College of IST can benefit from these courses (regardless of their technical and mathematical background).

This issue of ensuring accessibility is of particular relevance to instructors of the junior/senior level Informatics course on machine learning offered in the Applied Data Sciences major: DS 310 Machine Learning for Data Analytics (The Pennsylvania State University, 2023). This is because machine learning as a subject is inherently math-heavy, e.g., a lot of insights that power ML algorithms are derived from probability theory and linear algebra. Thus, it is impossible to do a deep-dive into any ML model or algorithm without showing them the mathematical assumptions and/or the derivations that are used to power the algorithms.

As an example, let us consider the case of support vector machines, a very popular machine learning model. A complete explanation of support vector machines would require the instructor to explain the formalism of the margin, use that to derive the optimization problem of the optimal margin classifier, and then show them how that optimization problem can be tweaked via the kernel trick that enables finding highly nonlinear margins (or decision boundaries) (Murphy, 2012). For a student who is unfamiliar with optimization problems, this math-heavy treatment of support vector machines can be quite daunting.

On the other end of the spectrum would be an instructor who treats support vector machines as a magical black-box model, i.e., instead of discussing with students how the model works internally, the instructor may choose to directly show students how to implement a support vector machine model inside their ML library of choice (e.g., scikit-learn (Scikit-learn developers, n.d.), Weka (University of Waikato, n.d.)) and use it for prediction/inference. While this treatment of support vector machines certainly makes the course accessible for all students, it robs mathematically inclined students of the chance to develop a deep understanding of how these models work and when can they be expected to work. I should note this way of teaching ML courses is followed by instructors on MOOC platforms (Coursera, n.d.; Udacity, 2022), where courses are mainly designed for accessibility. Thus, there is an inherent trade-off between a math-heavy treatment of the course and a more use-oriented treatment of the course.

First Thoughts on Course Design: Inspiration from Stanford CS 229: Machine Learning

As the author set about designing his course offering for Informatics students, the first point of reference was the Stanford CS 229: Machine Learning course (taught by Dr. Andrew Ng) (Stanford University, n.d.) as that was a course that the instructor had personally taken. The original CS 229 course is quite comprehensive in terms of topics covered, ranging from maximum likelihood estimation (MLE), linear and logistic regression, generalized linear models, Gaussian discriminant analysis, and support vector machines, followed by some basic lectures on neural networks and deep learning. However, there were a couple of reasons why this course was not a good fit for Informatics students at the College of IST. First, all ML topics covered in this course generally afforded their full mathematical treatment, which hinders the accessibility of course materials for a lot of Informatics students who do not have a strong mathematical background. Second, a lot of the algorithms covered in this class are not "essential" algorithms, e.g., in the views of the author, models like Gaussian discriminant analysis, generalized linear models, and expectation maximization are better suited to an advanced course offering on AI/ML.

Thoughts on Finalizing the Course Contents and Syllabus

Most introductory courses in machine learning across universities share a similar set of introductory ML models that belong to the classical ML paradigm: (i) linear regression; (ii) perceptron; (iii) logistic regression; (iv) decision trees; (v) random forests; (vi) bagging and boosting; (vii) ensemble models; (viii) support vector machines; and (ix) neural networks.

However, in the author's opinion, a key limitation of this "standard" and "conventional" ML syllabus is that it does not spend any classes discussing connections of these "conventional" classical ML models to statistical methods of learning. In the author's opinion, most students learn about statistical methods of learning such as maximum likelihood estimation (MLE) and maximum aposteriori estimation

(MAPE) (Murphy, 2012) in a general statistical estimation class, whereas in a standard ML class, they only learn about models like linear regression. The deep connection between MLE and linear regression (for example) is rarely explained to students inside a single class. The realization that most classical ML models like linear regression, ridge regression, lasso, etc. are all applications of the principle of MLE is a profound one, and the author really wanted to use MLE and MAPE as a starting point of discussion so that the probabilistic underpinnings of most conventional ML models can be laid bare.

As a result, the author decided to use the following syllabus for DS 310:

- The Principle of Learning Distributions from Data
- Maximum Likelihood Estimation/Max Aposteriori Estimation
- Principled approaches to supervised learning
  - Linear Regression
    - Ridge Regression
    - Lasso
  - Perceptron
  - Logistic Regression
  - Naïve Bayes Classifier
  - Decision Trees
  - Random Forests
  - The Principle of Bagging and Boosting
  - Ensemble Learning
  - Support Vector Machines
  - Introduction to Neural Networks/Deep Learning
- Feature Engineering
- Performance Assessment of Models
- Debugging ML Model Performance
- Relative Strengths and Weaknesses of Alternative Algorithms

Finalizing the Course Structure: Balancing Trade-Offs

As mentioned at the outset, the main priority for the course was to be accessible to all students without sacrificing the technical depth required to understanding the inner workings of the ML models and algorithms. The compromise that the author reached was by tweaking the evaluation criteria, i.e., the author made a conscious decision of describing and discussing each ML algorithm in four stages: (i) a high-level overview of what the algorithm is supposed to do and when should it be used (i.e., for regression or classification), a high level overview of the cost function used in the algorithm; (ii) a deep mathematical derivation of the cost function of the algorithm using statistical and probabilistic learning techniques, in many instances (this stage also covered the algorithm training procedure); (iii) a revisit of the high-level picture of the algorithm (this step was particularly important from an accessibility standpoint as the author felt that many students were left confused after seeing the deep mathematical treatment of the algorithm); and (iv) a programming-driven discussion around how to implement the model on real-world data and how to train it accurately.

Importantly, the author decided that no course evaluation would depend on the mathematical derivations, etc. Basically, the rationale behind this choice was the author's belief that in order to become a seasoned ML practitioner, it is essential to understand the deep mathematical underpinnings of these ML models. However, given the disparity in mathematical backgrounds, it is important to make sure that different students can still engage in learning during the class without being burdened by the fear of course evaluation depending on their grasp of the mathematical course content.

Most of the course evaluation was based around hands-on coding projects that required students to work with a real-world dataset and build an accurate ML model that could predict some desirable characteristic of individuals contained within that dataset.

Adaptations to Remote Learning Environments

I taught this course in the middle of the COVID-19 pandemic, and thus, I had to adapt my regular instruction style to suit a remote instructional setup. In my regular teaching style, I rely a lot on supplementing PowerPoint slides by writing things on the blackboard (e.g., math) if I need to explain something in greater detail. Since blackboards were no longer an option (because the webcam was not working), I decided to use a Microsoft Surface Pro to support my teaching. I rented a Microsoft Surface from the Pennsylvania State University so that I could write notes and math during my lecture (in order to replace the blackboard experience). I experimented with several virtual blackboard software, before finally deciding that the Zoom whiteboard feature was the most well-developed and most well-integrated with Zoom and hence had the most seamless performance. This was probably the first and most important adjustment that I had to make in order to transition to an online learning environment, without which it would have been really difficult for me to teach online. Next, the most challenging bit was the fact that all students in Zoom have their cameras turned off (and I did not feel comfortable asking people to turn on their cameras), so, adapting our teaching to their facial expressions is not possible. To mitigate this, I changed my instruction style, by spending around

10 min of time each class not teaching any material; instead during this time, I continuously encouraged students to ask any kind of questions that they might have. If they did not ask questions, I started asking questions to particular students who I knew were very proactive in class, and hence, they would be able to answer. This initial Q&A led to more students asking any questions that they might have had. I also spent a great deal of time repeating content or my words more than I usually would. All of these changes were greatly appreciated by students.

Reflections from Teaching DS 310: Machine Learning for Data Analytics

Overall, the course was well liked by students across three different semesters in which the DS 310 course was taught by the author. Students appreciated the high-level picture provided by the author during the discussion of every ML model/algorithm. For example, students (S1) and S2 mentioned:

> S1: "The professor always makes the teaching well prepared and interprets the essences and meanings of the difficult calculations very well, in an easily understood way. and the teaching content and modules are designed logically and reasonably, very easy to follow. I also very like the final research paper read session, which increases students' understanding and knowledge in terms of the application and trend of Machine Learning in current research fields."

> S2: "I'm taking another machine learning course in a different programming language, and the time he takes to talk through (1) the math, (2) the pseudocode, and (3) the actual implementation for every technique is incredibly helpful, and I wish my other professor did that. I like that he shared his slideshows and code samples for review after lecture, and the TA is always on top of Piazza posts for Q&A which is great."

Other students also found the "in-class examples and visualizations" (S3) very helpful in understanding the course material. Students also "really enjoyed the coding assignments as they learned so much more about machine learning" through them (S4).

The overall appreciation from the students is also evident by the teaching assistant's feedback:

> TA: "From the submission of the homework, most students exhibit good understanding of basic concepts about machine learning. Many of those homework questions are challenging and require deep understanding about concepts taught in the lecture. Still, most students successfully navigate these challenges."

> "Many students attended the TA office hours and asked insightful questions to further understand the contents in the lecture, even though some of these concepts go beyond the course requirements. This indicates students' dedication to learning but also the effectiveness of Prof. Yadav in fostering a curiosity for machine learning."

At the same time, it was clear to see that several students struggled with the course material, despite the authors' attempts at making the course more accessible. For example, a student (S5) remarked that "an hour of the class meeting should be devoted to explaining how to set up Jupyter notebooks and using scikit learn, since I have never really used Jupyter notebooks before." Other students (S6) also remarked that they found the "math presentation of the algorithms really boring and dry," and instead, they were "more interested in improving their ML based coding skills" (S7). These weaknesses are also noted by TA when engaging with students:

> TA: "However, the varying backgrounds of students have led to differing levels of proficiency, especially in programming. A small group of students failed to meet the expectation as they struggled with the coding part of the assignments. In particular, one student complained at my office hour that he/she did not understand how to use sci-kit learn because it is seldomly explained."

Another point of weakness mentioned in the course design was that all the hands-on coding projects were kept relatively open-ended, i.e., students were free to use the ML algorithm of their choice to implement a classifier that could generate accurate results. Unfortunately, the author decided that in order for the projects to be open-ended, it is important to cover all (or most) ML algorithms within the syllabus before students can be expected to start experimenting with their projects. This meant that the course evaluation timeline was bottom heavy, while the first half of the course relied on methodology-based homework assignments; the second half of the course was mainly focused on the implementation of the hands-on coding projects. Many students (S8, S9) felt that "the amount of work expected in the last two months of this class was just too much" and that instead, "it would be better if the amount of workload could have been distributed through the entire duration of the class." In a future iteration of the course, the author will act upon this feedback by releasing mini-projects at periodic intervals during the entire duration of the class.

Summary

Today's multifaceted societies widely adopt machine learning (ML) across various fields such as healthcare, criminal justice, and environmental protection. This surge in ML's popularity is paralleled by a growing eagerness among university students to master ML. Regrettably, the fundamental approach to ML within computer science programs is often dense with technical and mathematical content, deterring eager learners, especially those without a technical foundation. Information schools (iSchools), known for their embrace of various academic disciplines and commitment to multidisciplinarity, present a unique challenge: ML courses therein must accommodate students without extensive mathematical expertise. This paper illustrates the design and development of an ML course tailored for Informatics students at the Pennsylvania State University, aiming for accessibility. It explores a critical balance between a rigorous, math-intensive curriculum and a practical, application-focused approach. The paper illuminates the curriculum development decisions, assessment criteria, and teaching methods employed to achieve this balance, thus addressing the varied educational needs and interests of Informatics students.

References

Carnegie Mellon University. (n.d.). *B.S. in Artificial Intelligence*. Retrieved 5 December 5, 2023, from https://www.cs.cmu.edu/bs-in-artificial-intelligence/

Coursera. (n.d.). *Machine Learning Introduction [Specialization]*. Retrieved 5 December 5, 2023, from https://www.coursera.org/specializations/machine-learning-introduction

Georgia Tech. (n.d.). *Machine Learning*. Retrieved 5 December 5, 2023, from https://grad.gatech.edu/degree-programs/machine-learning

Murphy, K. P. (2012). *Machine learning: A probabilistic perspective*. MIT Press.

Penn State College of Information Sciences and Technology. (2023, December 5). *Wikipedia*. https://en.wikipedia.org/wiki/Penn_State_College_of_Information_Sciences_and_Technology

Penn State University. (n.d.-a). *Artificial Intelligence*. Retrieved 5 December 5, 2023, from https://bulletins.psu.edu/graduate/programs/majors/artificial-intelligence/

Penn State University. (n.d.-b). *Data Science, B.S. (Information Sciences and Technology)*. Retrieved 5 December 5, 2023, from https://bulletins.psu.edu/undergraduate/colleges/information-sciences-technology/data-sciences-bs/

Precedence Research. (2023). *Artificial Intelligence (AI) market*. Retrieved from https://www.precedenceresearch.com/artificial-intelligence-market

Scikit-learn developers. (n.d.). *Scikit-learn: Machine learning in python*. Retrieved from https://scikit-learn.org/stable/

Size, X. M. (2021). *Share & COVID-19 impact analysis*. By application (imaging and lighting, medical, satellite, electronics & semiconductors, and others (including R&D)), and regional forecast, 2020–2027. Available online: https://www.fortunebusinessinsights.com/xenon-market-101965. Accessed on 9 Feb 2021.

Stanford University. (n.d.). *CS229: Machine Learning*. Retrieved from https://cs229.stanford.edu/

The Pennsylvania State University. (2023). *DS 310: Machine Learning & Data Mining*. Retrieved from https://learning.ist.psu.edu/courses/ds/310

Udacity. (2022). *Introduction to machine learning*. Retrieved from 5 December 5, 2023, https://www.udacity.com/course/intro-to-machine-learning%2D%2Dud120

University of Michigan. (n.d.). *MS in Artificial Intelligence*. Retrieved 5 December 5, 2023, from https://umdearborn.edu/cecs/departments/computer-and-information-science/graduate-programs/ms-artificial-intelligence

University of Waikato. (n.d.). *Weka 3: data mining software in Java*. Retrieved from https://www.cs.waikato.ac.nz/ml/weka/

Unleashing the Problem-Solving Potential of Next-Generation Data Scientists

Lizhen Zhu and James Z. Wang

Abstract Data science, an emerging multidisciplinary field, resides at the intersection of computational sciences, statistical modeling, and domain-specific sciences. The current norm for data science education predominantly focuses on graduate programs, which presume a preexisting knowledge base in one or more relevant sciences. However, this framework often overlooks those who don't plan to pursue graduate studies, thereby limiting their exposure to this rapidly expanding field. Penn State addressed this gap by establishing one of the first undergraduate degree programs in data sciences, a collaboration between the College of Information Sciences and Technology, the Department of Computer Science and Engineering, and the Department of Statistics. One key component of this program is a project-focused, writing-intensive course designed for upper-class undergraduates. This course guides students through the entire data science project pipeline, from problem identification to solution presentation. It allows students to apply foundational data science principles to real-world problems, advancing their understanding through practical application. This chapter details the objectives, rationale, and course design, alongside reflections from our teaching experience. The insights provided could be helpful to instructors developing similar data science programs or courses at an undergraduate level, broadening the influence of this important field.

Keywords Undergraduate data science education · Project-based learning · Interdisciplinary curriculum · Interpersonal skills · Intrapersonal skills

Introduction

With the rapid advancements of digital sensing and database technologies, in today's world, data is ubiquitous, and its significance cannot be overemphasized. Data isn't

L. Zhu · J. Z. Wang (✉)
College of Information Sciences and Technology, The Pennsylvania State University, University Park, PA, USA
e-mail: ljz5180@psu.edu; jwang@ist.psu.edu

J. M. Carroll (ed.), *Innovative Practices in Teaching Information Sciences and Technology*, https://doi.org/10.1007/978-3-031-61290-9_8

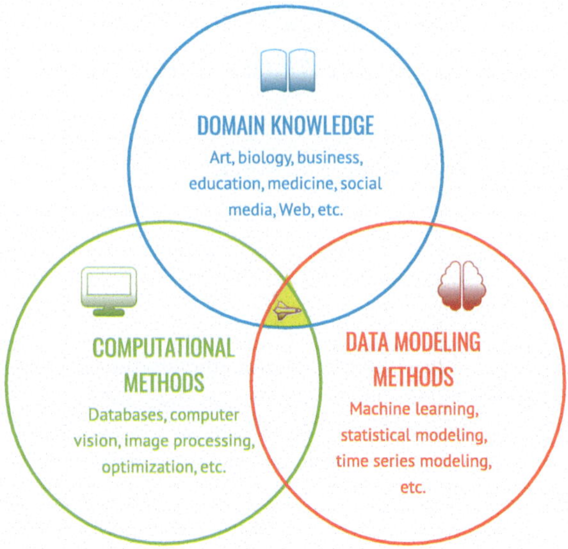

Fig. 1 Data science projects require domain knowledge, computational methods, and data modeling methods

just an integral part of our daily lives but is also used by scientists to discover new knowledge. A key challenge in leveraging the growing availability of data, however, lies in extracting meaningful insights from the ever-growing sea of information.

Data science is an emerging multidisciplinary field, crossing computational science, mathematical and statistical sciences, and application domain sciences that serves as a solution to this challenge. As depicted in Fig. 1, domain knowledge encompasses the understanding of the field of application, computational methods refer to the algorithms and processes used for data analysis, and data modeling methods consist of techniques for constructing and fine-tuning models. The intersection of these three components is where data science operates most effectively, leveraging domain expertise to apply appropriate computational and modeling methods, thereby driving valuable insights from data.

The demand for professionals with data science expertise has skyrocketed. Jobs in this sector are projected to increase by 36% from 2021 to 2031 (U.S. Bureau of Labor Statistics, 2023). However, because data science has to build on the foundational knowledge of these core fields, most training programs are at the graduate level, leaving undergraduate students without an opportunity to explore this exciting and rapidly expanding field.

At Penn State, we recognized this gap and created one of the first undergraduate data science degree programs in 2016 as an intercollege program between the College of Information Sciences and Technology, the Department of Computer Science and Engineering of the College of Engineering, and the Department of Statistics of the Eberly College of Science (Hallman, 2018). The data science major is structured into three options: computational, statistical modeling, and applied data sciences, each catering to different potential career trajectories.

The computational option emphasizes core computational skills, such as the design, implementation, and analysis of software that manages the volume, heterogeneity, and dynamic characteristics of large datasets and that leverages the computational power of modern hardware such as CPU and/or GPU cluster computers. The statistical modeling option focuses on statistical models and methods that are foundational to discovering and validating patterns in big data. The applied option equips students with the skills to integrate knowledge and techniques in problem-solving. Specifically, the emphasis is on addressing pressing scientific, organizational, and societal challenges.

Although students of the three options take different courses in their first two years, covering core methodologies relevant to their option, we begin to bring them into real-world problem-solving using data science in the third year through a required course, "applied data sciences," developed by James Wang, a coauthor of this chapter. Wang's diverse educational background includes advanced degrees in mathematics, computer science, and medical information sciences. His extensive experience in data science research spans a wide array of fields, including art, healthcare and medicine, plant biology, psychology, and meteorology.

This cornerstone course includes students from all three options and enables them to apply data science principles to real-world issues while sharpening their writing skills. This course is designed to provide students with a comprehensive understanding of data science by introducing fundamental principles through real-world examples and demonstrating how these principles apply to various methods and techniques commonly covered in data science courses. The course includes a term project and public speaking opportunities to enhance students' learning experiences. Upon completion, students should be able to proficiently navigate the entire process of a data science project, from problem formulation to data collection, technique selection, implementation, and performance evaluation. They will also be adept at effectively communicating actionable insights through written reports and oral presentations.

In this chapter, we will share our experiences in developing and teaching this project-focused, writing-intensive course. We will discuss the course design, the rationales behind its structure, and the insights we have gained from teaching it. We hope that our experiences could serve as a guide for instructors from other undergraduate data science programs looking to develop similar courses or programs.

Course Design

The course is structured to equip undergraduate students with the necessary knowledge and skills for executing data-driven projects in professional or research settings. After taking this course, students should master a comprehensive understanding of the data science pipeline. As illustrated in Fig. 2, a typical pipeline involves (1) problem formulation; (2) data acquisition; (3) data preparation, clean-

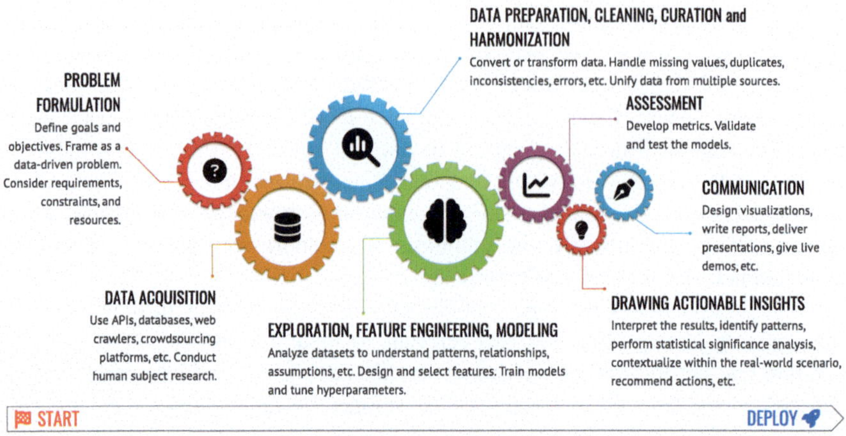

Fig. 2 Typical data science project pipeline. This diagram illustrates the sequential steps in a data science project, from problem formulation to communicating data science solutions to stakeholders

ing, curation, and harmonization; (4) exploration, feature engineering, and modeling; (5) assessment; (6) drawing actionable insights; and (7) communication.

To achieve this objective, the course is segmented into three key parts:

- **Foundation for Carrying Out Large-Scale Data Science Projects:** We include the concept of a data science project as an integration of computational methods, mathematical or statistical methods, and domain knowledge (Fig. 1). To motivate students about the importance of data science, we emphasize the view of data science as the fourth paradigm of conducting scientific exploration, following empirical science, theoretical science, and computational science (Fig. 3). Students are taught commonly used tools and are also equipped with nontechnical but vital skills like academic writing and presentation. Brief lectures from the instructor and teaching assistant (TA) on these topics are followed by simple in-class and take-home assignments to assess students' understanding. Furthermore, an in-depth analysis of the pipeline through a case study is used to reinforce knowledge.
- **Exposure to Important Data Science Subfields:** Based on their individual interests, students are assigned various topics to explore and present a talk summarizing the related literature. Seed papers provided by the instruction team aid the presenter in gathering additional materials. Basic literature search skills such as using various online databases, e.g., Google Scholar, IEEE Xplore, and ACM Digital Library, are covered by the instructor. During the student presentation, the presenter leads the discussion. The talk is peer-reviewed according to set criteria. Presenters are encouraged to promote audience participation. A subset of the seed papers also serves as reading assignments for all students. To ensure that students are keeping up with their reading, readiness assessment tests (RATs) are conducted in class based on the assigned reading. In addition

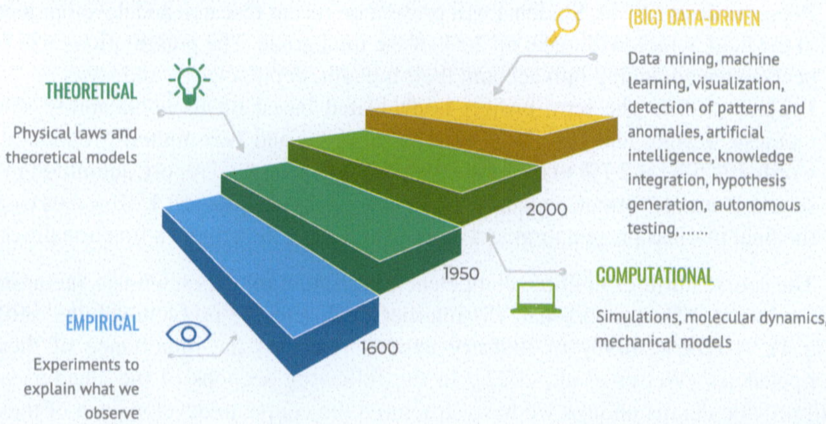

Fig. 3 The primary paradigms of scientific inquiry: empirical, theoretical, computational, and (big) data driven. Empirical science is rooted in observation and experimental data. Theoretical science focuses on the development of models and theories to explain phenomena. Computational science uses mathematical models and algorithms to solve complex problems. Finally, data-driven science utilizes sophisticated techniques to extract knowledge and insights directly from structured, semi-structured, and unstructured data

to this formal presentation assignment, the course also offers an opportunity for students to give an informal presentation on an interesting data science-related topic. The objective is to keep students up to date on the latest and most exciting developments in the field.

- **Real-World, Team-Based Project:** Students, in groups of two or three, work on a problem with real-world relevance. The problem, selected at the start of the semester in consultation with the course instructor and TA, should aim to utilize a large volume of complex digital data and statistical data modeling methods. Each team prepares a short one-to-two-page project proposal and an ignite talk (i.e., short, fast-paced presentation where the team shares their project ideas and approaches) at the beginning of the semester and receives writing and technical feedback for further enhancement. Regular meetings with the instructor and TA are set up to discuss project progress and provide detailed feedback on their reports. Toward the end of the semester, each team presents their project for peer review and submits an 8-to-10-page project report in a scholarly publication format for grading.

Student grades are composed of three elements.

1. **Participation** (30%): This part includes random class attendance checks and in-class activities, accounting for 15% of the grade. The other 15% is based on RATs. Over the semester, six RATs are administered, with the lowest score being disregarded.

2. **Presentations** (20%): Students will present on recent research and developments in the field, which will make up 20% of the final grade. The presentations will be peer-reviewed, taking into account their breadth, depth, clarity, and style.
3. **Project** (50%): The term project is evaluated based on its applicability, correctness, design, and functionality. The proposal and two midterm reports are worth 3%, 5%, and 7% to the grade, respectively. The final report, submitted in a scholarly article format, accounts for 20% of the grade. The last 15% comes from the final in-class presentation and demonstration of the project's functionalities.

The course nurtures multiple competencies essential for data scientists, including (1) technical, (2) intrapersonal, (3) interpersonal, and (4) problem-solving skills (Fig. 4). A recent survey of industry experts reaffirms the importance of these competencies (Weiser et al., 2022). In the following sections of the chapter, we will provide details on how we have structured the course to develop each of these skills in our students.

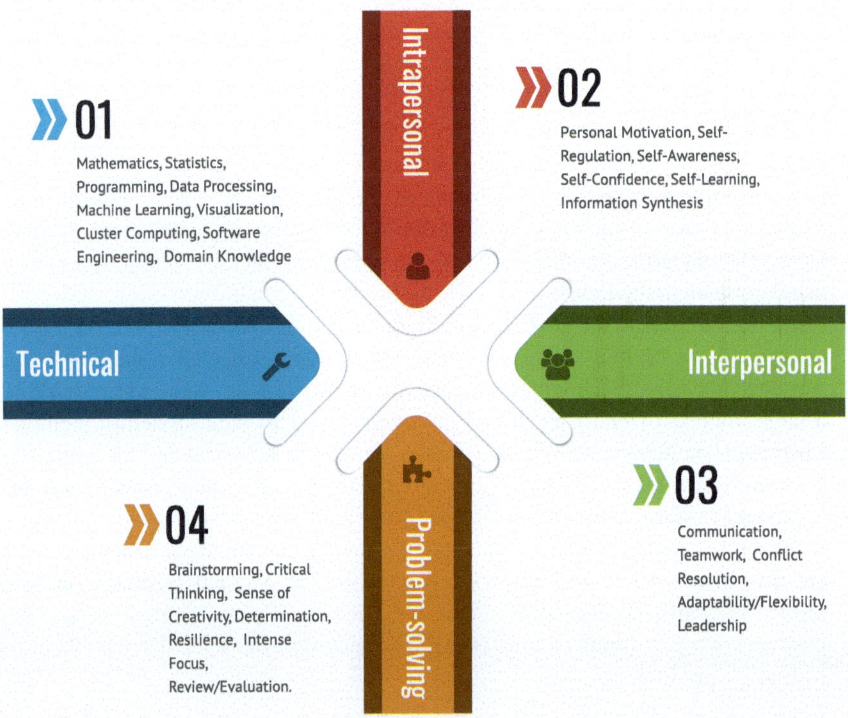

Fig. 4 An overview of the diverse skills targeted in the course curriculum, spanning technical, interpersonal, intrapersonal, and problem-solving capabilities, designed to equip students for success in their professional data science careers. Some of the skills can be categorized into multiple areas

Technical Skills: Continuous Learning, Case Analyses

This course provides an extension of the technical knowledge acquired in previous courses. There are mandatory prerequisites, including computer programming, calculus, probability, machine learning, data management, and privacy and security. Visual analytics is a suggested preparation. While previous experience with complex data types (e.g., multimodal/multimedia data, digital images, graphs) can be beneficial, it is not a requirement. During the initial phase of the course, we supplement students' existing knowledge through lectures on carefully chosen topics. These include the Message Passing Interface (MPI), PyTorch, cluster computing, cloud computing, and time series analysis.

Because data science is rapidly evolving with daily updates in methodologies and tools, successful data scientists require not just robust technical skills but also the ability to learn independently. Students need to develop the ability to source, comprehend, and use online scholarly articles, tutorials, textbooks, and short courses effectively. Furthermore, refining skills such as summarizing, explaining, discussing, presenting, critiquing, and defending scholarly work is an important aspect of their development in this field.

Continuous Learning In our pedagogical approach, we leverage *continuous learning* as a vehicle to not only reinforce students' cognitive grasp of the material but, more crucially, to provide them the opportunity to learn to learn, thereby equipping them with the adaptability and innovative thinking necessary for navigating future environmental shifts (Business Information Review Editorial, 2018). Continuous learning is characterized as a dynamic, unceasing cycle necessitating continuous modification and adaptation in response to feedback. Figure 5 depicts the cyclical process, which includes classroom instruction, assignments, and feedback.

The cycle begins with "classroom instruction," during which students acquire and process information, laying the groundwork for their knowledge base. The subsequent stage, "assignments," involves the integration and application of learned material, deepening comprehension, and sharpening skills. Finally, the "feedback" stage loops back into the cycle, providing critical insight into students' areas of strength and opportunities for growth. This reflective stage facilitates the refinement of both knowledge and skills, preparing the students for the next round of instruction and learning. The overarching objective of this methodology is the **holistic** development of students, promoting not only academic proficiency but also a lifelong passion for learning and self-improvement, critical attributes for a career in data science.

Each segment of this pedagogical cycle can adopt various formats. Assignments can take the form of in-class practical exercises, post-class homework, and presentations. In-class hands-on activities encourage students to replicate or make minor modifications to classroom demonstrations. This process empowers students to surmount perceived obstacles and fosters a willingness to experiment with novel concepts, such as coding with new software packages or formatting documents in unfamiliar LaTeX templates. Instructors can observe students during

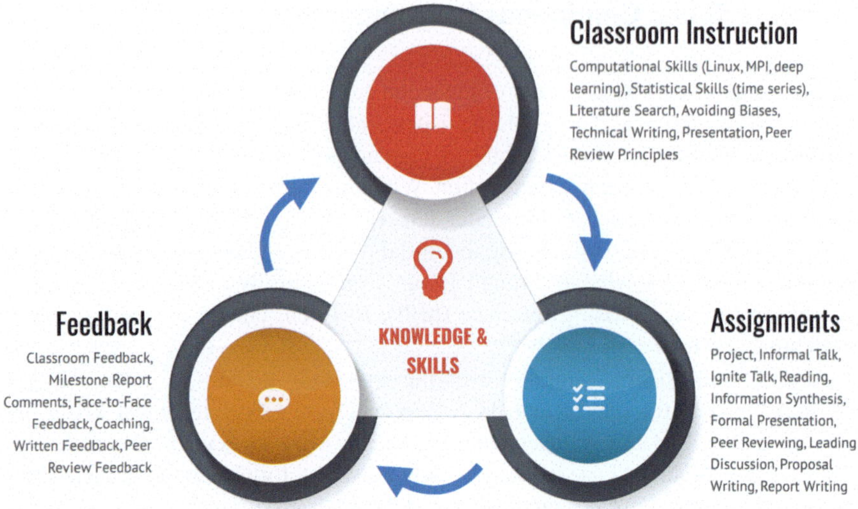

Fig. 5 This diagram encapsulates the cyclic and progressive nature of the learning process

these exercises, providing guidance to those facing difficulties and identifying common areas of misunderstanding for emphasis in subsequent instruction. Peer-to-peer interaction can also be encouraged to offer another form of feedback.

Homework serves as an opportunity for students to deepen their comprehension and requires more effort than in-class activities. Students are expected to engage in reflective thought or research to complete these assignments. As a form of feedback, grades alone may be insufficient. A feedback summary outlining common errors could be disseminated; even if some students do not commit these errors, it does not necessarily indicate comprehension. Additionally, brief, personalized comments can highlight specific mistakes in each student's work.

The design of subsequent course content can be influenced by students' demonstrated abilities. For student presentations, a peer-review process is often the most effective. The final project is crafted around this repetitive learning cycle, ensuring that our students remain on a path of consistent exploration and development.

Case Studies One crucial element we teach students in this course is the proper construction of a data science project pipeline, as depicted in Fig. 2. We introduce a case study at the semester's beginning to demonstrate the various stages involved in this process.

This case study revolves around the collaborative work of researchers (Li et al., 2012), spanning the fields of computing, statistics, and art history. This multidisciplinary team developed an innovative, data-driven approach to analyze Vincent van Gogh's distinct brushstroke styles. Through this research, they undertook a detailed statistical analysis of a broad set of features from automatically extracted brushstrokes. This analysis compared van Gogh's stylistic evolution over

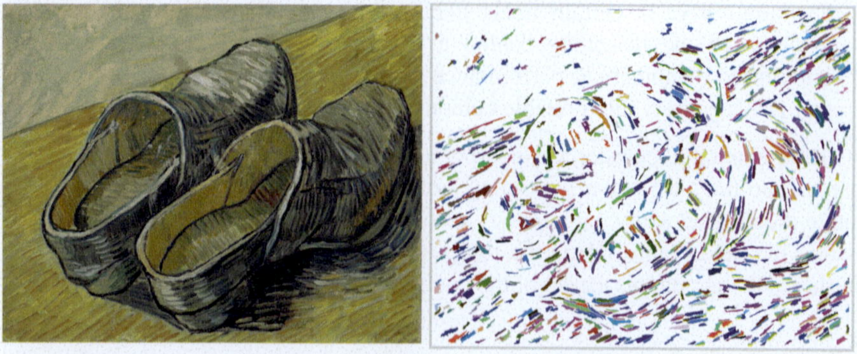

Fig. 6 Automatic brushstroke extraction for *A Pair of Leather Clogs* (Saint-Rémy-de-Provence, Autumn 1889, oil on canvas, 32.2 cm × 40.5 cm) by Vincent van Gogh (Li et al., 2012). Painting image courtesy of the Van Gogh Museum, Amsterdam (Vincent van Gogh Foundation). The brushstroke map is provided by the James Z. Wang Research Group (The Pennsylvania State University)

different periods and how it contrasts with the styles of his contemporaries. The team developed a new brushstroke extraction technique, fusing edge detection and clustering-based segmentation processes, to handle very high-resolution images with a large number of objects (i.e., brushstrokes). Figure 6 shows the automated brushstroke extraction result for one of the paintings.

The study's results revealed a pronounced rhythmic quality in van Gogh's brushstrokes. That is, van Gogh employed consistently shaped brushstrokes arranged closely together, thereby creating a patterned and recurrent impression. Interestingly, the characteristics that set apart van Gogh's paintings throughout various stages of his career are different from those that differentiate his works from those of his contemporaries.

We've chosen this as a notable case study due to its four distinctive characteristics:

- **Domain Knowledge:** The field of application, art history, is often perceived as separate from the disciplines of computing or mathematics. The success of this study underlines the significant role data science can play in addressing complex, real-world problems. It further illustrates the potential enhancements a data science project can gain from incorporating domain knowledge and closely collaborating with domain experts.
- **Bias Management:** The research highlights the crucial task of bias management at different stages of the data science pipeline. More discussions about bias in visual artwork analysis can be found in a recent scholarly paper (Zhang et al., 2022).
- **Innovative Approaches:** The solution to the problem required the invention of new methods rather than simply integrating existing tools. In particular, we stress to students the importance of appropriate problem-solving approaches when dealing with limited data.

- **Explainability:** In certain fields like art history and medicine, the capacity to elucidate the reasoning behind a model's decisions often outweighs the importance of numerical performance results. This work introduces a pipeline that yields entirely interpretable outcomes.

Throughout the semester, we incorporate a variety of recent data science research papers into the curriculum as reading assignments. This allows students to familiarize themselves with contemporary techniques, methodologies, and applications within the field. Table 1 lists the topics and example papers we assign. We will further elaborate on this aspect of the course in a subsequent section.

Table 1 Reading list for the course, with key topics and associated paper references. NLP: Natural language processing. AI: Artificial intelligence

Field	Topic	Papers
Machine learning	Unsupervised learning, self-supervised learning	Kingma and Welling (2014) and Chen et al. (2020)
	Few-shot learning, meta learning	Koch et al. (2015) and Finn et al. (2017)
	Reinforcement learning	Mnih et al. (2015)
	Adversarial machine learning	Moosavi-Dezfooli et al. (2016)
	Explainable AI	Lundberg and Lee (2017)
	Knowledge distillation	Hinton et al. (2015)
	Domain adaptation	Pan et al. (2010)
Computer vision	Object detection, semantic segmentation	Long et al. (2015) and Ren et al. (2015)
	Image generation, inpainting, style transfer	Gatys (2016) and Zhu et al. (2017)
	Human pose estimation, action recognition, facial recognition, head reconstruction, person reidentification	Schroff et al. (2015) and Cao et al. (2021)
NLP	Machine translation, sequence to sequence	Sutskever et al. (2014) and Cho et al. (2014)
	Language model	Vaswani et al. (2017) and Devlin et al. (2019)
Other applications	Image esthetics, computational photography, computer graphics, augmented reality	Datta et al. (2006) and Mildenhall et al. (2021)
	Recommendation system, music, speech, anomaly detection, fraud detection, robotics and autonomous driving	Sarwar et al. (2001) and Cadena et al. (2016)
	Healthcare, biology, astronomy, meteorology	Jumper et al. (2021) and Chen et al. (2014)
	Internet of things, finance, sports, privacy, law	Aletras et al. (2016) and Li et al. (2018)

Intrapersonal Skills: Self-Reflection, Personal Growth

The acquisition of relevant technical skills represents merely one facet of the knowledge needed for success in the field of data science. Equally critical are intrapersonal skills, comprising the individual's aptitude to comprehend and regulate their own emotional, motivational, and behavioral states. The development of these skills is closely tied to a student's background, leading to significant gaps in intrapersonal skill development across our highly diverse student population. Compounding this issue, some students may lack sufficient awareness of the importance of such skills. Against this challenging backdrop, we prioritize the fostering of intrapersonal skills that are directly related to problem-solving, facilitated through highly flexible task configurations.

Personal Motivation The inevitability of challenges in problem-solving underscores the need for personal motivation. This quality helps students set definitive goals, exhibit perseverance and resilience, and advance through self-guided learning. We use questionnaires to gather students' backgrounds and interests prior to course commencement. All topic choices for presentations, literature reviews, and projects are essentially student driven, reflecting their individual interests. This autonomy, however, presents a unique challenge: instructors must provide tailored guidance for each unique topic and ensure fair evaluation of assignments, irrespective of their subjects.

Self-Confidence Self-confidence is integral to the future success of data science students. Confidence emboldens students to assume risks and cultivate a positive self-perception, which is beneficial to their long-term well-being. We create a dynamic classroom environment and incorporate several presentation assignments coupled with question-and-answer sessions to encourage communication and opinion sharing. Our feedback approach is balanced, identifying areas of improvement while acknowledging and appreciating the students' efforts and progress.

Time Management and Organization Given the academic rigor faced by data science students, having to learn multiple interrelated fields, effective time management and organizational skills are paramount. Despite occasional student complaints about inadequate time to complete assignments and requests for deadline extensions, we strive to align assignment completion times with credit weighting. For extensive projects and literature review presentations, students' time management proficiency significantly impacts their performance. We emphasize the importance of effective time allocation at the project inception and maintain two milestone reports to monitor progress and provide timely reminders. To uphold the focus on learning, we accommodate assignment extensions, reinforcing the importance of thoroughness and quality in completed work. However, to ensure fairness, late submissions are subject to penalties.

Self-Reflection The capacity for self-reflection facilitates continuous progress. Students, through critical assessment of their strengths, weaknesses, and areas of improvement, can recalibrate their approach and foster ongoing development. This practice necessitates a higher degree of self-awareness and encourages an iterative, improvement-focused learning process. We ask students to include a thorough discussion of the merits and limitations of their approach in their project reports and presentations. Additionally, we advise them to remain cognizant of the various assumptions incorporated in their pipeline and to understand instances where such assumptions may not hold in the real world.

Interpersonal Skills: Effective Communication, Team Collaboration

In their daily work, data scientists frequently interact with a diverse array of stakeholders, including team members, domain experts, clients, senior management, and financial backers. Proficient interpersonal skills not only facilitate effective presentation of their findings and perspectives but also foster an enhanced understanding of other stakeholders' views, thereby sparking further exploration of specific topics.

Understanding Others The first step toward fruitful communication is to understand the perspectives of stakeholders from various fields and backgrounds. These stakeholders bring diverse views, methodologies, and feelings. In the term project, we strive to assign students from diverse backgrounds to the same team. This approach allows them to encounter varying viewpoints and learn to harness such diversity to their advantage. In our RATs, answers to most questions aren't directly available in the papers students read. Instead, they require students to grasp the principal ideas within these papers. Questions may include describing a method's pipeline or comparing the pros and cons of multiple methods. During student presentations, students in the audience are urged to engage by asking and answering questions. This way, they practice understanding each other through careful listening and resolve issues by reaching a consensus.

Being Persuasive Data science is not a field where mathematical proof is the norm. Yet, its methods must be useful in real-world scenarios. A data scientist needs to convey viewpoints clearly and understandably so others can appreciate their work. Adopting empathy, or putting oneself in the shoes of others, can make one's stance more acceptable. Being persuasive is, therefore, critical for data scientists. It enables them to communicate their findings and rationales effectively, secure support for their projects, and bridge the divide between technical and nontechnical audiences. This course helps develop this skill through the term project, where students work collaboratively with teammates to persuade the audience, through both writing and peer-reviewed talks that their problem formulation, approaches, and results are convincing.

Academic Writing Because this course is designated a "writing across the curriculum" course at Penn State, students are provided frequent and significant opportunities to write, revise, and discuss their writing. We introduce essential guidelines for proper formatting, citation, and plagiarism avoidance at the beginning of the course. We developed comprehensive grading rubrics to ensure clarity and fairness in evaluation. By incorporating the cycle illustrated in Fig. 5, we place an emphasis on written communication, aiming to ensure that student-generated reports are both comprehensible and professional. However, improving writing is a gradual process. Despite persistent feedback and correction, common issues such as citation format errors, grammatical and spelling mistakes, and the use of unprofessional visuals continue to be prevalent.

AI-Assisted Writing In the contemporary educational landscape, AI-based writing tools like ChatGPT and Grammarly have gained widespread popularity. While they can be a valuable asset in enhancing language fluency and readability, educators globally grapple with their proper use in writing-intensive courses. The foundational assumption remains—true learning necessitates comprehensive practice. This axiom, though rooted in the pre-digital age, retains its relevance even in today's AI-dominated era.

Notwithstanding, these AI tools have a unique advantage—they provide real-time feedback, which, when effectively utilized, can expedite the learning process. As a result, our course policy encourages students to use these tools, albeit with a caveat—the tools should be used strictly for enhancing language accuracy and readability and *not* for initial content generation. We underscore the significance of students developing their own writing skills and cultivating a distinctive "voice." Additionally, we mandate students to disclose the use of such AI assistance in their submitted writing assignments.

Oral Presentation We also develop students' oral presentation skills within the course structure, encompassing formats such as brief talks on current events in the data science and artificial intelligence field, literature discussions, ignite talks, and final project presentations. The inclusion of these elements allows for a comprehensive development of students' presentation abilities. Following multiple speaking engagements throughout the semester, the majority of students demonstrated an increased level of confidence and developed practical proficiency in professional public speaking.

Assessment To evaluate students' final reports, we have designed a comprehensive rubric, as presented in Table 2. For the literature presentation, we use a peer-evaluation form comprising four equally weighted dimensions: (1) breadth of topic coverage and organizational clarity, (2) depth of understanding as demonstrated in the presentation, (3) clarity of the delivery, and (4) presentation style, including multimedia usage, audience engagement, and overall interest level. For the final project presentation and demonstration, we resort to a similar peer-assessment methodology. It includes three equally weighted categories: (1) presentation, encompassing style, clarity, and accuracy, (2) creativity, measuring novelty in design, approach,

Table 2 Writing rubric for final term project report (total: 20 points)

	Needs improvement	Approaches expectations	Meets expectations	Excellent
Topic (4 pts.)	Poorly stated problem or unclear broader impact	Lacks clarity in problem or broader impact	Clear definition of problem and broader impact	Clear, well-articulated problem and broader impact. The value of the work is convincing
Outcomes (6 pts.)	Minimal description or lack of novelty in technical solution or result	Unclear description or inadequate justification of technical solution. Some effectiveness demonstrated. Lack of novelty	Clear description and justification of technical solution. Demonstrated effectiveness. Minimal novelty	Excellent discussion of technical solution and results. Good novelty. Exceptional insight into the solution's relevance to the problem. Strong evidence of effectiveness
Structure (6 pts.)	Fragmented and difficult to read. Does not provide a general picture of the project	Difficult to follow or verbose, repetitive descriptions	Easy to follow, clear descriptions, reasonably logical arguments	Clear, concise, easy to follow. Logical arguments. Appropriate section lengths
Grammar (2 pts.)	Unacceptable number of typos or grammatical errors	Noticeable typos or grammatical errors	Minimal typos or grammatical errors	Free from typos or grammatical errors
Citations (2 pts.)	Fails to use or incorrectly uses sources. Potential plagiarism will incur severe penalties	Uses sources but substitutes them for original ideas. Inconsistent citation format	Uses sources to support ideas. Appropriate quoting and paraphrasing but inconsistent citation format	Uses sources to support, not substitute for, original ideas. Skillful integration of published literature. Consistent citation format

or problem addressed, and (3) functionality/results, assessing the flawless operation of functions, the professional look and feel of the project, and the adequacy of the functions offered for the chosen data science application.

Problem-Solving Skills: Creative Thinking, Strategic Analysis

The ultimate objective of this course is to develop students' problem-solving capabilities. The complexity of real-world data science problems often lies in their multifaceted nature, characterized by the vastness of data, diversity in data types (including multimedia, sensor data, and graphs), imperfect data quality (such as noisy, incomplete, or inconsistent data), dynamic data attributes, and domain relevance. Additionally, data scientists have to continuously address various practical constraints or requirements, which may include appropriate modeling (for instance, avoiding overfitting or underfitting, parameter tuning), the interpretability/explainability of models, privacy and security considerations, ethical issues, and model maintenance tasks (like dynamically updating models as new data emerges).

Contrary to some other computing domains, such as numerical simulation and traditional databases, where a mathematical proof can confirm the correctness of an approach, the primary challenge in data science is the comprehensive orchestration of the entire pipeline. Therefore, identifying the optimal sequence of steps to constitute a valid pipeline is the focal point of this part of the course.

Creative Thinking Because of the complexity and nature of real-world data science problems, successful data scientists often need to devise novel solution pipelines. This underscores the value of pursuing data science as an undergraduate major, especially considering the abundance of open-source machine learning and data processing packages, coupled with the increasing sophistication of automated machine learning (AutoML) (He et al., 2021). We want to foster an environment where students can independently conceptualize and execute a complete pipeline, requiring minimal supervision from the course team.

This process starts with the *selection of project topics*. We encourage students to exercise their creativity by choosing their own project topics. While we are available to provide a "safety net" by suggesting topics, it is noteworthy that nearly all teams can formulate their own interesting ideas. We have communicated our expectations to them clearly: suitable project topics must integrate computational methods, mathematical/statistical approaches, and domain knowledge. Additionally, these topics should exhibit sufficient complexity and scale, have real-world implications, and feature some innovative aspect(s).

The following is a sample of twenty project topics conceived by our students:

- News topic analysis based on text data
- Modeling feature representations for human emotional states
- Color histogram matching approach for speech emotion recognition
- High-energy particle tracking in CERN detectors

- Decoding neuron interactions: predicting brain activity from semantic stimuli using fMRI
- Utilizing growth rings and PDSI to predict climatic events
- Natural language analysis of Twitter for the 2020 US Presidential Election
- Predicting Major League Baseball pitcher injuries with Statcast data
- Classification of maritime vessels using feature engineering
- Analysis of the intersection between mental health, nutrition, and happiness
- A deep learning approach for malware classification
- Sports moneyline odds prediction using machine learning
- Analysis of homeless youth data to predict drug usage
- Automated identification of Spotted Lanternflies (Lycorma delicatula)
- Computer vision-based yoga posture recognition and rating system
- Speech-to-speech translation with adaptable text to speech
- Multimodal sarcasm detection in sitcoms
- Using explainable AI to interpret complex breast cancer detection models
- Transient detection in "nearby" galaxies using TESS data-driven pipeline
- Detect early-onset Parkinson's disease

This array of project topics covers several interdisciplinary fields including natural language processing (NLP) and text analysis, emotional and psychological analytics, speech and audio processing, neuroscience and biomedical studies, physics and astronomy, climate science and dendrochronology, sports analytics, cybersecurity, maritime science, social studies, entomology, and health and fitness.

Throughout the term project, we underscore the importance of novelty and creativity. We adopt an *apprenticeship* approach in our coaching, rather than the conventional production pipeline model used in typical lecture-based courses. We mandate students to incorporate an innovation segment in their project and to argue for that in their project reports, thus motivating them to continually evaluate the novelty of their project. While we provide necessary guidance and relevant academic papers when they need them, the responsibility is largely on the students to brainstorm their own solutions as a team.

When presenting their project, we also encourage students to be creative. They need to devise strategies to engage their peers while ensuring the inclusion of all necessary content within a limited amount of time. Students have used animations, self-recorded videos, and real-life demos, among other techniques.

Strategic Analysis The field of data science is currently in a state of explosive development, with new methods and challenges constantly emerging, often superseding their predecessors. To keep pace with these ongoing advancements, having refined academic acumen and insight is important. Data scientists need to have the capability to identify impactful R&D topics, discern datasets with wide-ranging potential benefits, and recognize methods with enduring value.

Developing such an insightful perspective is not an overnight task. In our course, we facilitate the nurturing of this outlook by discussing the motivations behind some classic projects. We provide students with "seed" academic papers to help initiate their literature reviews and projects, thereby intentionally training them in strategic analysis.

Furthermore, we incorporate an exemplary real-world industrial project in the discussion to expose students to the strategic analysis done by leading data scientists in the industry when they face challenging problems. Specifically, we let students watch a video talk by Dr. Dhiraj Joshi of IBM Research, summarizing his work related to emotion understanding from videos (ISTatPENNSTATE, 2018). Students are then tasked with summarizing three key takeaways from both the talk's technical content and its delivery. According to their feedback, they have learned strategic analysis skills and formed new perspectives on some classic learning techniques such as the support vector machine (SVM) (Cortes and Vapnik, 1995), once they understood their practical applications. Some students expressed doubts about the extent of information one could extract from a dataset, while others generated ideas on evaluating the success of an AI system. The students also gleaned professional presentation skills from the video, including the presenter's style, use of gestures, incorporation of video content, and smooth transitions between topics.

Decision-Making Efficient and rational decision-making constitutes a vital component in the personal and professional growth of data science students. Data science projects often necessitate extensive resource management, spanning time, financial investment, data, computational resources, and manpower. Upon performing strategic analysis during problem selection, data gathering, and data processing phases, it is necessary to make informed decisions that maximize positive outcomes and mitigate risks.

This skill is developed during the entirety of the term project. Students are tasked with not only determining their primary topics but also the strategies employed to overcome project challenges, leveraging the resources available to them. In a data science pipeline, a decision is involved in every step, and these decisions can have an impact on the quality and validity of the work. Furthermore, students are invited to pose questions during in-person project meetings with the instruction team, scheduled twice in the semester. We offer guidance to ensure their progress when they encounter hesitations or face problematic decisions. This approach progressively expands students' competency in effective decision-making.

Table 3 provides a summary of the contributions of individual course components to the development of each of the four major skills. While our course facilitates the development of diverse skills, there remains room for reflection and improvement. The subsequent section will address these aspects from both the student and instructor perspectives.

Reflections and Lessons Learned

Student Feedback A news article initially covered the course's introduction, featuring a few student comments as testimonials (Hallman, 2018). Among them

Table 3 Development of diverse skills through various components of the course

Component	Skills			
	Technical	Intrapersonal	Interpersonal	Problem-solving
Short presentation	Emerging technologies	Self-confidence, Personal motivation	Public speaking	Information collection
MPI, PyTorch, Cluster/Cloud computing	UNIX, programming			
Time series	Statistical method			
Writing lectures and Workshop			Technical writing	Research and Information collection
Industrial invited speaker talk	Industrial applications of technologies		Business communication	Problem solving, Critical thinking
Literature review	Domain knowledge, Machine learning	Self-learning, Self-confidence, Personal motivation	Scholarly communication	Research skills, Analytical thinking
Readiness assessment tests	Domain knowledge, Machine learning			Project pipeline, Analytical thinking
Ignite talk			Public speaking, Self-confidence	
Project proposal			Team collaboration, Proposal writing	Idea generation, Creativity
Project and Reports	Data analytics, Visualization, Domain knowledge	Self-reflection, Personal motivation, Time management	Team management, Leadership	Critical thinking, Research skills
Peer review activities			Feedback and Communication	Assessment and Evaluation

was a statement from Ryan Jaeger, who was a second-year student when he took the course. Jaeger articulated his understanding of the course's objectives, stating, "In my mind, a good data scientist is able to manipulate data and perform the machine learning modeling necessary to answer a given question. However, a great data scientist is also able to generate meaningful research questions, place their research in context, and clearly communicate the significance and novelty of their work to many audiences. I believe the emphasis on communication in DS 340W will help me to become a great data scientist."

We consistently gather student feedback and course ratings at the end of each semester to continuously refine the course experience. Overall, students have expressed satisfaction and enthusiasm about the course. For instance, in the fall semester of 2022 (following the COVID-19 pandemic, but prior to the introduction of ChatGPT), the authors taught the course and received positive ratings: the course achieved a median score of 6 out of 7, while the instructor received a full score of 7 out of 7. These scores were based on responses from 72% of the class. Written feedback from students further revealed various aspects of the course that positively influenced their learning experience:

- **Hands-on Experience:** Students valued the practical, applied nature of the course, which offered firsthand experience in data science.
- **Reviewing Literature:** Analyzing articles and other people's methods helped students understand current industry practices and facilitated the application of these methods in other academic settings.
- **Freedom and Creativity:** Students appreciated the liberty to choose their own topics for open-ended projects, which made the learning process more enjoyable and enriching.
- **Focus on Presentation Skills:** The course not only developed data science skills but also nurtured presentation abilities. Despite initial fear, students recognized the importance of public speaking and appreciated the focus on presentation skills.
- **Writing Workshops:** Workshops and one-on-one feedback from the professor and TA were deemed very helpful. Students learned academic writing techniques specific to data science.
- **Use of Tools:** Tools like Overleaf for writing formal reports were beneficial for students' learning.
- **Literature Talk:** Activities like literature talk were appreciated for boosting confidence in public speaking.
- **Professor's Enthusiasm:** The professor's passion for the subject matter was noted as a key factor that made the class more engaging and information more retainable.
- **Class Attendance and Interesting Talks:** Regular attendance and exposure to intriguing talks made the learning process more effective and enjoyable.

The students suggested improvements for enhancing their learning experience, including the following:

- **RATs:** Some students suggested that RATs should be replaced with class discussions, where every student has an opportunity to participate. Some students found the grading standards too stringent.
- **LaTeX:** While acknowledging the usefulness of the Overleaf tool, several students found it challenging to use. They proposed holding workshops or dedicated classes to teach students how to use Overleaf effectively.

- **Project:** Some students suggested replacing the single, semester-long project with multiple smaller projects. This would offer more opportunities for presentations and writing, thus enhancing their learning.
- **Programming:** Students asked for more programming assignments to help them learn useful technologies. Some students found the supercomputer assignments challenging and suggested more demonstrations and hands-on practice.
- **Writing:** A few students found the feedback on papers and presentations to be vague and requested more detailed and constructive feedback.

Successes and Challenges From the instructors' perspective, the course has achieved its goals in several key areas. It has successfully instilled in students the significance of data science, guided them through the typical project pipeline, enabled them to comprehend the challenges involved in implementing each pipeline step correctly, and cultivated their problem-solving skills for handling large and complex real-world problems. It has also developed their communication abilities. We are pleased that most students can complete a project using a robust methodology and articulate their findings both in writing and orally.

The course's effectiveness and impact are further substantiated by the feedback we've received from past alumni of the program. They've noted that this is one of the few courses where the acquired skills can be directly applied in a professional setting as a data scientist.

However, in spite of our continuous efforts to enhance the course, certain challenges persist such as the following:

- **Balancing Topics:** It is difficult to cover so many different topics in one semester. We dedicate a substantial portion of class time to student presentations to provide them with opportunities to present formally and informally in front of a live audience. This inevitably reduces the time for lectures or hands-on programming, which is reflected in the subpar performance of some students in programming assignments.
- **Teamwork Issues:** Some students, overwhelmed with multiple courses, struggle to contribute equally to team projects. As a result, we often encounter issues such as uneven contributions or students only taking part in tasks they are already proficient in.
- **Plagiarism:** We use Turnitin to detect plagiarism in final reports. Every semester, despite we emphasize academic integrity repeatedly, we can detect plagiarizing. With AI-based writing tools like ChatGPT and paraphrasing tools becoming more available, plagiarism detection becomes even more challenging. Our current approach allows students to use these tools to enhance readability and language quality, but they must disclose such use. Furthermore, using ideas from published work without appropriate attribution or citation is deemed plagiarism. However, it's important to note that detecting this type of plagiarism can be extremely difficult.
- **Artistic Sense:** The development of esthetically pleasing visualization and talk slides is a crucial aspect of communication skills in data science. However,

nurturing an artistic sense in students within a short semester proves to be a daunting task. This is reflected in their reports and presentation slides. We have encouraged them to learn from professionals by reading high-quality published articles and adopting professional slide templates, but observable improvements have been minimal over the semester. We believe data science students should be encouraged to engage more with artworks, enroll in art classes, practice creating art, and/or analyze artworks to gradually develop their artistic sense.

- **Risk-Taking and Novelty:** We encourage students to embrace risk-taking and innovation when selecting their project topics. However, our observation is that most teams tend to be more conservative. This cautious approach might be attributed to this being their first experience with real-world problems. It may require multiple such exposures before they can gain more confidence and venture into unexplored territories.
- **Insufficient Scale or Complexity:** Although we provide students with access to supercomputers and occasionally commercial cloud platforms, only a handful of project teams managed to work with data of a scale and complexity that justified the use of such resources. Consequently, many students did not significantly enhance their skills in managing large-scale, complex problems through this course.
- **Fair Peer Reviewing:** Generally, students are fair and constructive in evaluating their classmates' presentations. However, we have occasionally observed variations in scoring where some students consistently provide low scores, while others give exceptionally high scores. We attribute these discrepancies not to any malicious intent but rather to the fact that different individuals have varying criteria for assessing others' work.

Concluding Remarks

The process of teaching this course has been a continual learning experience. It has refined our understanding of effective pedagogy for teaching data science, guided by evolving trends in the field, emerging educational methodologies, and most importantly, the feedback and progress of the students themselves. Designing and implementing a project-focused, writing-intensive data science course that nurtures an impactful and engaging learning experience requires a careful balance of theoretical knowledge and practical application, individual growth and collaborative engagement, and adhering to established norms and promoting innovation. Seeing our students undertake intensive explorations of complex concepts, apply what they've learned to real-world challenges, and steadily grow into emergent data scientists has been an enormously gratifying affirmation of our teaching methodology's effectiveness.

Acknowledgments We extend our appreciation to Jia Li, who codeveloped and co-instructed the course in its formative years. We also thank Tongan Cai, Yukun Chen, Dolzodmaa Davaasuren, Yu Li, Benjamin Wortman, and Sitao Zhang, who have contributed to the course development and/or delivery in their capacity as teaching assistants. The course used the Extreme Science and Engineering Discovery Environment, which was supported by the National Science Foundation (NSF) under Grant No. ACI-1548562, the Advanced Cyberinfrastructure Coordination Ecosystem: Services & Support (ACCESS) Program supported by NSF under Grant Nos. OAC-2138259, OAC-2138286, OAC-2138307, OAC-2137603, and OAC-2138296, and the Amazon Web Services through the Educate Program. The data science research of the authors is supported by the NSF under Grant No. 2216127. The data science research of Wang is also supported by the NSF under Grant Nos. 2234195, 2205004, and 2015943, the National Institutes of Health under Grant No. R01EB030130, and the National Endowment for the Humanities under Grant No. HAA-287938-22. Wang would also like to acknowledge the Amazon Research Awards program for the generous gifts (2019-2023).

References

Aletras, N., Tsarapatsanis, D., Preoţiuc-Pietro, D., & Lampos, V. (2016). Predicting judicial decisions of the European court of human rights: A natural language processing perspective. *PeerJ Computer Science, 2*, e93.

Business Information Review Editorial (2018). The importance of continuous learning for innovation, progression and survival. *Business Information Review, 35*(1), 6–8.

Cadena, C., Carlone, L., Carrillo, H., Latif, Y., Scaramuzza, D., Neira, J., Reid, I., & Leonard, J. J. (2016). Past, present, and future of simultaneous localization and mapping: Toward the robust-perception age. *IEEE Transactions on Robotics, 32*(6), 1309–1332.

Cao, Z., Hidalgo, G., Simon, T., Wei, S.-E., & Sheikh, Y. (2021). OpenPose: Realtime multi-person 2d pose estimation using part affinity fields. *IEEE Transactions on Pattern Analysis and Machine Intelligence, 43*(1), 172–186.

Chen, T., Kornblith, S., Norouzi, M., & Hinton, G. (2020). A simple framework for contrastive learning of visual representations. In *Proceedings of the International Conference on Machine Learning* (pp. 1597–1607). PMLR.

Chen, Y., Lin, Z., Zhao, X., Wang, G., & Gu, Y. (2014). Deep learning-based classification of hyperspectral data. *IEEE Journal of Selected Topics in Applied Earth Observations and Remote Sensing, 7*(6), 2094–2107.

Cho, K., Van Merriënboer, B., Gulcehre, C., Bahdanau, D., Bougares, F., Schwenk, H., & Bengio, Y. (2014). Learning phrase representations using RNN encoder-decoder for statistical machine translation. In *Proceedings of the Conference on Empirical Methods Natural Language Processing* (pp. 1724–1734).

Cortes, C., & Vapnik, V. (1995). Support-vector networks. *Machine Learning, 20*, 273–297.

Datta, R., Joshi, D., Li, J., & Wang, J. Z. (2006). Studying aesthetics in photographic images using a computational approach. In *Proceedings of the European Conference on Computer Vision, Part III* (pp. 288–301). Springer.

Devlin, J., Chang, M.-W., Lee, K., & Toutanova, K. (2019). BERT: Pre-training of deep bidirectional transformers for language understanding. In *Proceedings of the Conference of the North American Chapter of the Association for Computational Linguistics: Human Language Technologies* (Vol. 1, pp. 4171–4186).

Finn, C., Abbeel, P., & Levine, S. (2017). Model-agnostic meta-learning for fast adaptation of deep networks. In *Proceedings of the International Conference on Machine Learning* (pp. 1126–1135). PMLR.

Gatys, L. A., Ecker, A. S., & Bethge, M. (2016). Image style transfer using convolutional neural networks. In *Proceedings of the IEEE Conference on Computer Vision and Pattern Recognition* (pp. 2414–2423).

Hallman, J. (2018). Data sciences program blends interdisciplinary training for a growing industry. *Penn State News*.

He, X., Zhao, K., & Chu, X. (2021). Automl: A survey of the state-of-the-art. *Knowledge-Based Systems, 212*, 106622.

Hinton, G., Vinyals, O., & Dean, J. (2015). Distilling the knowledge in a neural network. In *Proceedings of NIPS Deep Learning and Representation Learning Workshop*.

ISTatPENNSTATE (2018). *Penn State Startup Week 2018 - Dr. Dhiraj Joshi, Senior Research Scientist, IBM Watson*. https://www.youtube.com/watch?v=j5r0vaXQCr8

Jumper, J., Evans, R., Pritzel, A., Green, T., Figurnov, M., Ronneberger, O., Tunyasuvunakool, K., Bates, R., Žídek, A., Potapenko, A., et al. (2021). Highly accurate protein structure prediction with AlphaFold. *Nature, 596*(7873), 583–589.

Kingma, D. P., & Welling, M. (2014). Auto-encoding variational Bayes. In *Proceedings of the International Conference on Learning Representations*.

Koch, G., Zemel, R., Salakhutdinov, R., et al. (2015). Siamese neural networks for one-shot image recognition. In *Proceedings of the Deep Learning Workshop, in Conjunction with the International Conference on Machine Learning* (Vol. 2). Lille.

Li, H., Ota, K., & Dong, M. (2018). Learning IoT in edge: Deep learning for the internet of things with edge computing. *IEEE Network, 32*(1), 96–101.

Li, J., Yao, L., Hendriks, E., & Wang, J. Z. (2012). Rhythmic brushstrokes distinguish van Gogh from his contemporaries: Findings via automated brushstroke extraction. *IEEE Transactions on Pattern Analysis and Machine Intelligence, 34*(6), 1159–1176.

Long, J., Shelhamer, E., & Darrell, T. (2015). Fully convolutional networks for semantic segmentation. In *Proceedings of the IEEE Conference on Computer Vision and Pattern Recognition* (pp. 3431–3440).

Lundberg, S. M., & Lee, S.-I. (2017). A unified approach to interpreting model predictions. In *Advances in Neural Information Processing Systems, 30*.

Mildenhall, B., Srinivasan, P. P., Tancik, M., Barron, J. T., Ramamoorthi, R., & Ng, R. (2021). NeRF: Representing scenes as neural radiance fields for view synthesis. *Communications of the ACM, 65*(1), 99–106.

Mnih, V., Kavukcuoglu, K., Silver, D., Rusu, A. A., Veness, J., Bellemare, M. G., Graves, A., Riedmiller, M., Fidjeland, A. K., Ostrovski, G., et al. (2015). Human-level control through deep reinforcement learning. *Nature, 518*(7540), 529–533.

Moosavi-Dezfooli, S.-M., Fawzi, A., & Frossard, P. (2016). DeepFool: A simple and accurate method to fool deep neural networks. In *Proceedings of the IEEE Conference on Computer Vision and Pattern Recognition* (pp. 2574–2582).

Pan, S. J., Tsang, I. W., Kwok, J. T., & Yang, Q. (2010). Domain adaptation via transfer component analysis. *IEEE Transactions on Neural Networks, 22*(2), 199–210.

Ren, S., He, K., Girshick, R., & Sun, J. (2015). Faster R-CNN: Towards real-time object detection with region proposal networks. In *Advances in Neural Information Processing Systems, 28*.

Sarwar, B., Karypis, G., Konstan, J., & Riedl, J. (2001). Item-based collaborative filtering recommendation algorithms. In *Proceedings of the 10th International Conference on World Wide Web* (pp. 285–295).

Schroff, F., Kalenichenko, D., & Philbin, J. (2015). FaceNet: A unified embedding for face recognition and clustering. In *Proceedings of the IEEE Conference on Computer Vision and Pattern Recognition* (pp. 815–823).

Sutskever, I., Vinyals, O., & Le, Q. V. (2014). Sequence to sequence learning with neural networks. In *Advances in Neural Information Processing Systems, 27*.

U.S. Bureau of Labor Statistics (2023). *Data Scientists*. https://www.bls.gov/ooh/math/data-scientists.htm. [Online; accessed 14-June-2023].

Vaswani, A., Shazeer, N., Parmar, N., Uszkoreit, J., Jones, L., Gomez, A. N., Kaiser, Ł., & Polosukhin, I. (2017). Attention is all you need. In *Advances in Neural Information Processing Systems, 30.*

Weiser, O., Kalman, Y. M., Kent, C., & Ravid, G. (2022). 65 competencies: Which ones should your data analytics experts have? *Communications of the ACM, 65*(3), 58–66.

Zhang, Z., Li, J., Stork, D. G., Mansfield, E., Russell, J., Adams, C., & Wang, J. Z. (2022). Reducing bias in ai-based analysis of visual artworks. *IEEE BITS the Information Theory Magazine, 2*(1), 36–48.

Zhu, J.-Y., Park, T., Isola, P., & Efros, A. A. (2017). Unpaired image-to-image translation using cycle-consistent adversarial networks. In *Proceedings of the IEEE International Conference on Computer Vision* (pp. 2223–2232).

Providing Iterative Cybersecurity Hands-On Learning Experience: Reflections on Teaching Cyber Defense to Second Year IT Students

Peng Liu

Abstract Although hands-on cybersecurity labs are being increasingly adopted in higher education institutes, we observe that the learning experiences gained by entry-level students are mixed. This chapter describes the author's reflections on and analysis of why the students' learning experiences are mixed. This chapter also describes the author's efforts to improve the students' learning experiences through an iterative hands-on lab design approach. Compared to widely adopted hands-on security labs, the proposed approach retrofits the connections between the labs in a sequence in such a way that the students can gain the ability to master a few advanced hands-on cyber defense skills. The description of the proposed approach includes the authors' reflections on the relation between the unique cybersecurity education challenges and the design principles of iterative hands-on labs, the likely pitfalls in implementing iterative labs, and the main challenges in assessing the extent to which the students' learning experiences are improved. The abovementioned reflections are based on the author's teaching of CYBER 262 (Cyber-Defense Studio) at Penn State University.

Keywords Hands-on cyber-defense skills · Iterative teaching labs · Teaching students in Cybersecurity major

Introduction

During the past two decades, an increasing number of bachelor's degree programs dedicated to cybersecurity have been established in higher education institutions in the United States. Since the primary goal of these degree programs is to produce cybersecurity professionals, the students are directly introduced to the essential concepts and practices in the field of cybersecurity during their first or second

---

P. Liu (✉)
College of Information Sciences and Technology, Pennsylvania State University, University Park, PA, USA
e-mail: pxl20@psu.edu

J. M. Carroll (ed.), *Innovative Practices in Teaching Information Sciences and Technology*, https://doi.org/10.1007/978-3-031-61290-9_9

year of study. As a result, the introductory cyber defense courses in these degree programs and the introductory cyber defense courses offered in a CS (computer science) or ECE (electrical and computer engineering) degree program have very different kinds of contents. In particular, the main differences are as follows:

- When teaching a fundamentally important cybersecurity concept such as buffer overflow attack, the students from a CS/ECE program would have already taken a computer architecture course in which the related computer system concepts, including buffers, stack frames, and segmentation faults, are systematically taught. In contrast, the students from a cybersecurity program are most likely a person who has never studied such computer system concepts. Since the lab instruction author (implicitly) assumes that the students know the computer systems concepts, the lab instructions usually will not explain such concepts again to the students. As a result, the lab instructions can become very difficult for the students from a cybersecurity program to comprehend.
- Taking the buffer overflow attack as an example, the lab instructions usually involve a number of Linux command line commands used by a student to compile the given vulnerable source code, reconfigure the memory protection (e.g., non-executable stack) of the operating systems, and so forth. Although the students from a CS/ECE program have already gain a substantial amount of hands-on experience on the command line commands through their previous computer system courses, the students from a cybersecurity program are likely a person who is new to these command line commands.
- Taking a buffer overflow attack lab as an example, assuming that the students are familiar with the computer system concepts involved in the lab, the lab designer will usually let the learning objectives (e.g., a set advanced hands-on cyber defense skills) be built on top of the computer system concepts. While these learning objectives are reachable for the students from a CS/ECE program, they are in general far away from being reachable for the students from a cybersecurity program: The technical depth is clearly beyond what a student can handle.

Due to these differences, the hands-on labs designed as part of an introductory cyber defense course offered in a CS/ECE degree program **simply cannot be reused** in a cybersecurity degree program. Accordingly, the development of a hands-on lab portfolio, which consists of a set of teaching labs, becomes a major challenge in teaching entry-level courses in bachelor's degree programs dedicated to cybersecurity.

In the remaining of this book chapter, we first summarize the main limitations of the existing hands-on lab portfolios in cybersecurity degree programs. Second, we analyze the root causes of the limitations and propose an idea to effectively address the limitations. Third, we present the design and implementation of a newly proposed hands-on lab portfolio which is used in the author's teaching of CYBER 262 (Cyber-Defense Studio) at Penn State University.

Background and Motivation

The Hands-On Lab Portfolio Designed for Entry-Level Cybersecurity Courses Usually Consists of a Set of Labs Loosely Related to Each Other

Taking the entry-level courses in the B.S. Cybersecurity degree program at Penn State University, for example, four typical hands-on lab portfolios are as follows. The first hands-on lab portfolio consists of the following labs:

- The dictionary attack lab allows the students to gain hands-on experience on assessing the strengths of passwords and breaking into a password-protected computer.
- The AES (Advanced Encryption Standard) data encryption lab allows the students to gain hands-on experience on encrypting/decrypting sensitive data files and key management (e.g., key generation and distribution).
- The Firewall ACL (Access Control List) configuration lab allows the students to gain hands-on experience on configuring and testing a basic type of firewalls.
- The Buffer Overflow lab allows the students to gain hands-on experience on exploiting memory corruption bugs.
- The Penetration Testing lab allows the students to use commodity tools such as Kali Linux[1] to conduct preliminary penetration testing.

The second hands-on lab portfolio consists of the following labs: The first four labs are the same as the first portfolio, but the last lab is replaced by the Snort lab which allows the students to gain hands-on experience on using Snort,[2] an open-source intrusion detection system (IDS) and adding new manually written detection rules to the IDS.

The third hands-on lab portfolio consists of the following labs: The five labs and are the same as the first portfolio except that the fourth lab is replaced by the aforementioned Snort lab.

The fourth hands-on lab portfolio consists of the following labs: The five labs and are the same as the first portfolio except that the second lab is replaced by the Wireshark lab[3] which allows the students to gain hands-on experience on capturing and interactively browsing (and analyzing) the cyberattack traffic running on a computer network.

Although the four typical hands-on lab portfolios are not identical, it is not very difficult to find that they have the following common characteristics:

---

[1] https://www.kali.org/

[2] https://www.snort.org/

[3] https://www.wireshark.org/faq.html

- Characteristic 1: All the lab portfolios intend to allow the students to gain hands-on experiences in multiple subfields of cybersecurity. For example, the first lab portfolio covers five subfields: user authentication, data encryption, network security, endpoint memory corruption attack, and penetration testing, respectively.
- Characteristic 2: The set of covered subfields are often loosely related to each other. For example, the hands-on experiences gained in the subfield of data encryption are not directly useful when the students are doing a hands-on lab in the field of penetration testing.
- Characteristic 3: For each subfield the lab portfolios intend to cover, using only one lab is offered to the students.

Unfortunately, these common characteristics make the lab portfolios not very effective in helping the students achieve their learning objectives. In particular, the limited effectiveness could be summarized as follows. First, since only one lab is offered to the students in each subfield, and as a result in case the involved cybersecurity concept has significant technical depth, the "one-shot" learning experience would be far away from reaching the depth reached by an entry-level cybersecurity analyst. Second, when the set of labs in one portfolio are only loosely related to each other, the learning experience gained by a student in the earlier weeks of the semester would not play an essential role in helping the student gain a different level of technical depth in the later weeks of the semester. As a result, even the last lab in the semester could not allow the student to experience the depth reached by an entry-level cybersecurity analyst.

How Could the Hands-On Lab Portfolio Designed for Introductory Cybersecurity Courses Be Further Improved?

The limited effectiveness of the existing hands-on lab portfolios motivates us to achieve a better trade-off between the breadth and technical depth of a lab portfolio. In particular, **our idea** is as follows: On the one hand, for the cybersecurity concepts that simultaneously have fundamental importance (in the discipline) and well-earned reputation for being notoriously difficult, an iterative hands-on lab design approach should be taken to retrofit the connections between the labs in a sequence in such a way that the students can gain the ability to master a few advanced hands-on cyber defense skills. On the other hand, in order to nurture the students' broad interests, introductory videos and demos can be assigned to the students to replace the corresponding depth-lacking hands-on labs.

Overview of the Proposed Hands-On Lab Portfolio

In order to implement the aforementioned idea, the proposed hands-on lab portfolio has the following properties:

- Each set of labs are supposed to be focused on one fundamentally important cybersecurity concept (and its extensions).
- Each set of labs are supposed to be offered in a per-determined serial order.
  - The serial order enables the lab portfolio to function as an *iterative* learning project which consists of exclusively of a sequence of iterations. From the perspective of helping the students fully understand the cybersecurity concept, the purpose of each iteration is to provide the students with one more layer of understanding on top of what the students understand through the previous iterations.
- Each lab shall explicitly specify the following preconditions and postconditions:
  - Precondition A: What understanding (about the involved cybersecurity concept) a student should have before starting his or her work on the teaching lab?
  - Precondition B: What hands-on skills a student should have before starting his or her work on the lab?
  - Postcondition A: What new understanding a student should gain through his or her work on the lab?
  - Precondition B: What new hands-on skills a student should gain through his or her work on the lab?
- When Lab Y is the next lab of Lab X in the aforementioned serial order, the following relationships must hold:
  - Postcondition A of Lab X must play an essential role in meeting precondition A of Lab Y.
  - Postcondition B of Lab X must play an essential role in meeting precondition B of Lab Y.
- Every lab in the portfolio must not be too challenging for a student who meets all of the preconditions of the lab. On the other hand, no single Lab X is easy for a student who does not yet meet the preconditions of the next lab of X in the aforementioned serial order.
- A student who does not meet the preconditions of the second lab in the aforementioned serial order but succeeds in completing every lab in the portfolio should gain much deeper understanding (and hands-on skills) about the involved cybersecurity concept than what he or she gains through the first lab in the serial order.

Designing the Proposed Hands-On Lab Portfolio

The proposed hands-on lab portfolio for CYBER 262 at Penn State University mainly consists of five particular hands-on labs. The serial order (among the labs) is as follows: The first lab is the "simplest buffer overflow" lab; the second lab is the "using a piece of shellcode to overwrite a variable" lab; the third lab is the "using a calculator to predict the address of the shellcode" lab; the fourth lab is the "Return to Libc" lab; the final lab is the "format string vulnerability" lab. Below, we will present the five labs one by one.

The Simplified Buffer Overflow Lab

This lab has the following characteristics:

- Precondition A: The students know nothing about the buffer overflow attack concept.
- Precondition B: The students have preliminary experience on Linux machine command line commands through their learning experience in CYBER 100 (Computer Systems Literacy).
- Postcondition A: The students gain the first layer of understanding about the concept, that is, an unusually long input can overwrite an irreverent variable stored in the memory chip.
- Postcondition B: The students master the following hands-on skills: (1) They can use GDB, a debugging tool, to "see" the variable (and its value) in the memory chip. (2) They can also print the variable's address.
- The lab employs the simplest possible buffer overflow incident to make sure that the lab will not be too challenging for the students. In particular, (a) the C program used in the lab has only ten lines of code; (b) the C program is pre-compiled into an executable file which the students can directly use in the alb.

The "Using a Piece of Shellcode to Overwrite a Variable" Lab

This lab has the following characteristics:

- Precondition A: This precondition is the first postcondition of the simplified buffer overflow lab.
- Precondition B: This precondition is the second postcondition of the simplified buffer overflow lab.
- Postcondition A: The students gain one more layer of understanding about the concept, that is, the stack can hold machine code.

- Postcondition B: The students master the following hands-on skills: With the professor-provided source code, the students can use `gcc`, a compiler, to generate a piece of shellcode.
- The lab employs the simplest possible shellcode to make sure that the lab will not be too challenging for the students. Instead of teaching the shellcode through disassembly code, this lab provides the students with the corresponding source code which is much easier to comprehend.

The "Using a Calculator to Predict the Address of the Shellcode" Lab

This lab has the following characteristics:

- Precondition A: This precondition is the first postcondition of the shellcode lab.
- Precondition B: This precondition is the second postcondition of the shellcode lab.
- Postcondition A: The students gain one more layer of understanding about the concept, that is, the attacker can infer post-infection memory layout.
- Postcondition B: The students master the following hands-on skills: Isaac Newton can predict how objects orbit; the students can predict the address of injected malicious code.
- The lab employs an online "decimal digits to hexadecimal digit" converter to make sure that the lab will not be too challenging for the students. Instead of asking the students to do calculations on hex numbers, the students can use do calculations on decimal numbers.

The First Milestone Achieved Through the Sequence of Labs So Far

- *Milestone 1*: After the students also overwrite the "Return Address" field with the predicted address, a magical moment arrives: the shellcode is automatically executed!

The Return-to-Libc Attack Lab

This lab has the following characteristics:

- Precondition A: This precondition is the first postcondition of the address calculation lab.

- Precondition B: This precondition is the second postcondition of the address calculation lab. The address prediction technique used in the preceding lab will be reused in this lab.
- Postcondition A: The students gain one more layer of understanding about the concept, that is, the existing code in a computer can be abused to do a lot of harm.
- Postcondition B: The students master the following hands-on skills: The students can overwrite the "Return Address" field with the address of a particular Libc function and put the corresponding input arguments in the right locations in the stack.
- The lab employs only one Libc function to make sure that the lab will not be too challenging for the students.

The Second Milestone Achieved Through the Sequence of Labs So Far

- *Milestone* 2: Without injecting any malicious code, the students can trigger the computer to jump to the existing code.

The Format String Vulnerability Lab

This lab has the following characteristics:

- Precondition A: This precondition is the union of the first postconditions of the previous four labs.
- Precondition B: This precondition is the union of the second postcondition of the previous four labs.
- Postcondition A: The students gain one more layer of understanding about the concept, that is, specially crafted format strings can be leveraged to modify the content in an arbitrary memory location.
- Postcondition B: The students master the following hands-on skills: The students can craft special format strings to read or modify the content in a target memory location.
- The lab reuses several techniques learned in the previous labs to make sure that the lab will not be too challenging for the students.

The Third Milestone Achieved Through the Sequence of Labs

- *Milestone 3*: The students can use malicious `printf()` format strings to modify the content in an arbitrary memory location. Then the students can reuse the techniques learned in the previous labs to launch code injection attack.

Implementing and Assessing the Proposed Hands-On Lab Portfolio

Implementation. The proposed hands-on lab portfolio is implemented through reusing a subset of the elements (e.g., vulnerable C programs, command line commands) used in the following three SEED labs[4] which are widely used in educating students in CS/ECE degree programs:

- The Buffer Overflow Vulnerability Lab
- The Return-to-Libc Attack Lab
- The Format String Vulnerability Lab.

Preliminary Assessment. In order to evaluate the effectiveness of the proposed lab portfolio, we ask the following research questions. (RQ1) Is the first lab too challenging for the students to achieve their learning objectives? (RQ2) Is any of the five labs too easy for the students? (RQ3) Is any of the five labs too difficult for the students? (RQ4) Can the students master certain advanced hands-on skills (e.g., the skills used by a professional cybersecurity analyst)?

Regarding RQ1, if the average score of the students who did the first lab is fairly low, one could conclude that the first lab is too challenging. Accordingly, we examined the grades of the students (in total, 34 students) who did the first lab in the Fall 2021 semester. We found that the average score is 83. In addition, only three students received a grade lower than 80. They received grades of 65, 60, and 60, respectively. Regarding RQ2, we calculated the average grades for each lab and found that the average grade of every lab is below 90. Regarding RQ3, we found that the average grade of every lab is higher than 80. Regarding RQ4, we observed that majority of the students mastered the following fairly advanced hands-on skills: The students can predict (via a calculator) the address of injected malicious code; the students can overwrite the "Return Address" field with the address of a particular Libc function and put the corresponding input arguments in the right locations in the stack; the students can use malicious `printf()` format strings to modify the content in an arbitrary memory location.

[4] https://seedsecuritylabs.org/

Concluding Remarks

Motivated by the limitations of the existing hands-on labs used in entry-level courses in a B.S. Cybersecurity degree program in which the set of labs are loosely related to each other, we propose a different way to design such lab portfolios: for the cybersecurity concepts that simultaneously have fundamental importance and well-earned reputation for being notoriously difficult, an iterative hands-on lab design approach should be taken to retrofit the connections between the labs in a sequence in such a way that the students can gain the ability to master a few advanced hands-on cyber defense skills.

In this chapter, we present the design and implementation of a specific lab portfolio (for CYBER 262 at Penn State University) which adopts the iterative hands-on lab design approach. In addition, we assess the effectiveness of the newly developed lab portfolio and report the lessons learned.

Learning About Explainable AI with Very Little Programming

For 4th Year Undergrads (or Younger)

Jonathan Dodge, Antoinette Diawuo, Enyan Dai, and Rupak Das

Abstract Teaching students about explainable AI is emerging as a critical topic for industry readiness, but doing so is difficult. This is because it contains a vast array of concepts, each of which may be new to students. In this chapter, we will go through the strategies we have honed to make this course accessible to fledgling data scientists. Along the way, we try to make very few assumptions about programming ability, mathematical knowledge, or human-computer interaction experience in an effort to provide a broadly accessible course.

Keywords Explainability · AI · HCI · Teaching Information Science and Technology · Low-code

Class Overview

As of this writing, we are now offering a course for the third time on explainable artificial intelligence (DS402: XAI), which is the main character of our story. In spring 2023, it wore a different hat as IST402, where it debuted with Antoinette Diawuo as classroom assistant, and we taught 14 students. In fall 2023, we moved it over to data science with Enyan Dai assisting, where it has attracted more students at 36. In spring 2024, with Rupak Das assisting, it grew even more, up to 47 students.

J. Dodge (✉) · A. Diawuo · E. Dai · R. Das
College of Information Sciences and Technology, The Pennsylvania State University, University Park, PA, USA
e-mail: jxd6067@psu.edu

J. M. Carroll (ed.), *Innovative Practices in Teaching Information Sciences and Technology*, https://doi.org/10.1007/978-3-031-61290-9_10

Cataclysm Sentence

Explainable artificial intelligence (XAI) is composed of approaches that researchers use to **create** explanations, such as feature attribution or training data distribution information, and **evaluate** explanations, such as in-lab user studies or heuristic evaluation.[a]

[a]NPR Radiolab (2020) hosted a podcast in which they repeated Dr. Richard Feynman's thought experiment with artists and a variety of other people. This thought experiment asks: *"If in a certain cataclysm, where all of the scientific knowledge is to be destroyed, but only one sentence is to be passed down to the next generations of creatures, what would be the best thing? The thing that contains the most information in the least number of words?"*.

What is XAI?

Broadly speaking, XAI sits at the nexus of human-computer interaction (HCI) and artificial intelligence (AI). In addition to these two main pillars, it involves a variety of other disciplines as well. Examples include visualization, analytics, psychology, philosophy of science, and so on. As such, we have readings from a very wide range of venues, from hardcore ML/AI (NeurIPS, ICML) to human-computer interaction (ACM CHI), combinations of AI+HCI (ACM IUI, ACM TIIS), robotics (IROS), computer vision (CVPR), and visualization (VAST). See Table 1 for the full reading list.

XAI is and will remain an important topic to study because AI-infused systems are already pervasive in many humans lives, but most people do not have much background in the underlying mathematical concepts that power such systems. As such, XAI techniques will exert a strong influence in determining whether laypersons *understand the capabilities and limitations* of AI-infused systems. Such understanding will drive usage to be more appropriate (i.e., people use the system when it works and decline to use the system when it does not work, as described in Schemmer et al., 2023). When the first author discusses this research topic with family members and friends who do not have a technical background, this characterization of XAI suffices: *Suppose you are developing something like a self-driving car and wish to sign a document from a regulator indicating that "We have tested this and it is not a danger to society." XAI contains whatever tools and techniques one might use to make such an assertion.*

Table 1 Current reading list for the course. Several of these papers feature prominently in the classwork, whether using a software package (as in #19 and #22) or directly working with the code akin to that underpinning the research work (as in #13)

#	Authors. (Year). Title. Venue
1	Abdul et al. (2018). Trends and trajectories for explainable, accountable, and intelligible systems: An HCI research agenda. *Proceedings of the 2018 CHI Conference on Human Factors in Computing Systems*
2	Silver et al. (2018). A general reinforcement learning algorithm that masters chess, shogi, and Go through self-play. *Science*
3	Penney et al. (2021). The Shoutcasters, the Game Enthusiasts, and the AI: Foraging for Explanations of Real-Time Strategy Players. *ACM Trans. Interact. Intell. Syst.*
4	Vinyals et al. (2019). Grandmaster level in StarCraft II using multi-agent reinforcement learning. *Nature*
5	Kim et al. (2016). Evaluation of StarCraft Artificial Intelligence Competition Bots by Experienced Human Players. *ACM CHI Conference Extended Abstracts*
6	Kulesza et al. (2015). Principles of explanatory debugging to personalize interactive machine learning. *ACM International Conference on Intelligent User Interfaces*
7	Huang et al. (2018). Establishing Appropriate Trust via Critical States. *IEEE/RSJ International Conference on Intelligent Robots and Systems (IROS)*
8	Dodge et al. (2022). How Do People Rank Multiple Mutant Agents? *27th International Conference on Intelligent User Interfaces*
9	Lim et al. (2009). Why and Why Not Explanations Improve the Intelligibility of Context-Aware Intelligent Systems. *2009 SIGCHI Conference on Human Factors in Computing Systems*
10	Goodfellow and Papernot (2017). The challenge of verification and testing of machine learning. *Blog post*
11	Hoffman (2018). Metrics for Explainable AI: Challenges and Prospects. *arXiv*
12	Binns et al. (2018). "It's Reducing a Human Being to a Percentage": Perceptions of Justice in Algorithmic Decisions. *Proceedings of the 2018 CHI Conference on Human Factors in Computing Systems*
13	Dodge et al. (2019). Explaining Models: An Empirical Study of How Explanations Impact Fairness Judgment. *Proceedings of the 24th International Conference on Intelligent User Interfaces*
14	Amershi et al. (2019). Guidelines for Human-AI Interaction. *Proceedings of the 2019 CHI Conference on Human Factors in Computing Systems*
15	Vaccaro et al. (2018). The Illusion of Control: Placebo Effects of Control Settings. *Proceedings of the 2018 CHI Conference on Human Factors in Computing Systems*
16	Mitchell et al. (2019). Model Cards for Model Reporting. *Proceedings of the Conference on Fairness, Accountability, and Transparency*
17	Jin et al. (2022). EUCA: the End-User-Centered Explainable AI Framework. *arXiv*
18	Eiband et al. (2021). How to Support Users in Understanding Intelligent Systems? Structuring the Discussion. *26th International Conference on Intelligent User Interfaces*
19	Ribeiro (2016). "Why Should I Trust You?": Explaining the Predictions of Any Classifier. *Proceedings of the 22nd ACM SIGKDD International Conference on Knowledge Discovery and Data Mining*
20	Anderson et al. (2020). Mental Models of Mere Mortals with Explanations of Reinforcement Learning. *ACM Trans. Interact. Intell. Syst.*

(continued)

Table 1 (continued)

#	Authors. (Year). Title. Venue
21	Poursabzi-Sangdeh et al. (2021). Manipulating and Measuring Model Interpretability. *Proceedings of the 2021 CHI Conference on Human Factors in Computing Systems*
22	Lundberg et al. (2017). A unified approach to interpreting model predictions. *Proceedings of the 31st International Conference on Neural Information Processing Systems*
23	Weiss et al. (2018). Extracting Automata from Recurrent Neural Networks Using Queries and Counterexamples. *Proceedings of the 35th International Conference on Machine Learning*
24	Groce et al. (2014). You Are the Only Possible Oracle: Effective Test Selection for End Users of Interactive Machine Learning Systems. *IEEE Transactions on Software Engineering*
25	Eykholt et al. (2018). Robust Physical-World Attacks on Deep Learning Visual Classification. *2018 IEEE/CVF Conference on Computer Vision and Pattern Recognition*
26	Kahng et al. (2017). ActiVis: Visual Exploration of Industry-Scale Deep Neural Network Models. *IEEE Transactions on Visualization and Computer Graphics*
27	Chen et al. (2018). AnchorViz: Facilitating Classifier Error Discovery through Interactive Semantic Data Exploration. *23rd International Conference on Intelligent User Interfaces*

XAI Often Focuses on Laypersons

In our view, an XAI researcher is uniquely positioned to provide commentary on *educating* laypersons about AI, because conducting user studies on this topic requires providing a mini-lecture about our AI-infused system[1] before participants use it. In fact, several of our studies (e.g., Anderson et al., 2020) have *excluded* those with a technical background to get a better understanding of layperson attitudes and abilities with respect to AI-infused systems. Part of why our research has focused on laypersons is inspired by the ACM Code of Ethics, which states: *... all people are stakeholders in computing* (ACM, 2024). Further, the recent White House Blueprint attempts to empower such stakeholders: *... You should know how and why an outcome impacting you was determined by an automated system...* (White House, 2022). The layperson focus has paid dividends in the class. A number of sophomores and juniors have ended up taking the class because the prerequisites are minimal. All that said, none of the authors are specifically trained in pedagogy, so some of the pitfalls we found may be obvious to some readers.

[1] Definition from Amershi et al. (2019): *"In this paper we use **AI-infused systems** to refer to systems that have features harnessing AI capabilities that are directly exposed to the end user."*.

Challenges of XAI

Ultimately, the main challenge of XAI is that the AI-infused systems we want to explain are complex, but we wish to present simple information to the user. However, in the process of simplification, we do not wish to lose soundness. One of the key insights offered by Kulesza et al. (2015) is to be *sound* and *complete* but do not *overwhelm*. On the periphery of this main challenge, we have others, such as "How do I measure the quality of my explanation process or artifact?". Hoffman (2018) offer two main mechanisms, termed *satisfaction* and *goodness*, which are akin to empirical and analytical methods, respectively. For a satisfaction-based study, we first operationalize a task, then measure the performance of participants with our novel explanation, and compare it to some alternative. On the other hand, a goodness-based study essentially appeals to expert opinion (e.g., heuristic evaluation, as in Amershi et al., 2019). Note that measurement in these manners begets yet more challenges, because now we need to construct a corpus of suitable tasks (satisfaction) or strategies to collect expert opinion (goodness).

First Author's Class Philosophy

Throw a lot of weird pitches to the students, if they can hit any one of them, that is good enough to pass the course. If they can hit multiple of them, they should consider grad school!

My reasoning for employing such a philosophy is that very few HCI-oriented students take sufficient AI so as to understand everything the course covers, while very few DS-oriented students take sufficient HCI so as to get everything. As such, I view the class as an exercise in diverging thought to let the students explore their interests in an independent way, whether that means they do a deep dive on one class concept or a shallow dive on many class concepts. One downstream consequence of this philosophy is that students may complete their classwork on VERY different timelines based on time available, how quickly they reach an aha moment, programming ability, which concepts they focus upon, and so on. Thus, to help support students who find the scope of this course daunting, I offer flexible deadlines, repeatedly encouraging students to *submit work they have done well instead of quickly*.

Course Content

We teach the course primarily through paper presentations from a curated reading list (there are very few lectures), with each student being responsible to contribute toward presenting one paper. Another large chunk of time goes toward in-class labs, which the students are responsible to finish as homework. In addition to these two graded elements, we have a final project report, a final project presentation, and participation credit.

Though the course division does not perfectly mirror this, it covers five main topics, which are AI, HCI, explanation strategies, evaluating AI and explanations, and loose ends. The current offering is designed to fit a twice-weekly meeting schedule, meaning 30 class periods, each 75 minutes long. We divide each session roughly in half, giving a 5 minute break between halves.

Paper Presentations (10%)

Paper presentations are intended to teach students a strategy the first author learned as a PhD student from Margaret Burnett (Oregon State University) in which the presenter marks up the paper directly and uses it as a presentation artifact. The main goals of this approach are fourfold:

1. Presenting via the paper does not consume a lot of preparation time for the presenter, as compared to preparing slides.
2. After the presentation, everyone in the audience has approximately "read" the paper.
3. We avoid requiring a textbook, which means the class has no cost beyond tuition.
4. We train students to learn in the same way top-tier researchers do, by **engaging directly with primary sources**.

Early in the term, the professor leads all the presentations to both teach students the contents of the paper and also demonstrate the presentation approach to clarify expectations. After having seen a few presentations, the professor requires each student to sign up to present one paper to the rest of the class, and anything that no one signs up to present becomes the teaching team's task to present. During each student presentation, the professor is available to clarify and interject content that the students would have no particular reason to know. As an example, during the presentation on ModelCards (Mitchell et al., 2019), the professor points out how this work later inspired ServiceCards (Philomin & Hallinan, 2022) because many of Amazon's enterprise AI-infused systems use a *combination* of learned components, as opposed to just a single model.

As the reading list in Table 1 illustrates, the papers we cover in the class come from a very broad set of venues, which allows students to more effectively choose their own adventure, aligning with their interests and abilities.

Labs (50%)

With most of the labs, the main goal is to get students hands-on experience working with code that implements AI and XAI techniques. The labs offer students an opportunity to work with the code and ask questions of the teaching team as needed. We design the labs to take longer than the class time available, so they will need to work on them a bit more as homework. We drop the lowest lab score to give students a better chance to recover if they get sick or stumble. The full listing of labs is available in Table 2.

The first series of six labs is based on sequential decision-making in Markov decision processes (MDPs) using an object-oriented Python program. MDPs are a way to model transitions between states based on chosen actions that allows for nondeterminism. Figure 1 shows a very small toy MDP that we study in the class, though MDPs can model a very broad class of problems. One specific problem we study is the "parking domain" from one of Alan Fern's classes at Oregon State, highlighted in Figs. 2 and 3.

In each successive lab, we incorporate a new type of agent, such as Q-learning or policy iteration (see Fig. 4 for more on these topics). In some labs, we ask them to do a very small amount of programming (e.g., to build a simple probabilistic policy to use as a baseline), but most labs involve exploring correct implementations that the teaching team provides. In particular, the goal is frequently to understand how parameters of the agent and MDP affect performance. As an example, the students adjust the learning rate of a Q-learning agent, create learning curves, and reason about their charts.

The seventh lab is standalone, designed to help the students brainstorm research questions (RQs) and begin finding their own sources for the final project. It consists of a series of guided activities, starting with determining a topic that may interest them. Next, we show the students the types of RQ from a taxonomy by Shull et al. (2007). Then, the students generate a few RQs and discuss what they have in small groups. Next, we teach them about the concept of a "win-win" RQ, which has the property of providing something interesting to say regardless of which way the data comes out (i.e., if the RQ assumes my novel method will outperform state of the art, do I have a paper if it does not?). At this point, the students revise their RQs to make them win-win. For the last step, we show them another taxonomy from the same source that demonstrates a number of user study methodologies. Then, the students pick a few study methodologies that are well-suited to answer their RQ. For homework, the students find three pieces of related work, updating their RQ as necessary if some other researchers have already answered it convincingly, as an example.

The second series of two labs switches gears to use notebooks instead of an integrated development environment (IDE), as well as switching over to studying fairness. The code we examine in these labs is based heavily on work the first author performed while an intern at IBM AI Research, which culminated in an award-winning conference paper (Dodge et al., 2019). This series of labs focuses on the

Table 2 Labs contained within the course

#	Title: *(Focus)*. Description
1	**Getting familiar with Markov decision processes (MDPs)**: *(Domain understanding)*. Understand the problem structure through pictures. Begin working with a random agent that can record results obtained by taking actions and observing rewards
2	**How to measure and create probabilistic policies:** *(Domain understanding)*. Develop a handcrafted baseline policy. Develop more intuition about how one can use knowledge of the structure of the MDP to do better than random, as the probabilistic policies will serve as a baseline for the learners we will study later
3	**Getting started with reinforcement learning**: *(Light AI programming)*. Understand and measure the quality of a Q-learning agent, which is a value-based technique. See how the Q-value table updates with each action-observation pair. Explore the impact of different parameter settings for the agent and the MDP on the learning performance
4	**Training a deep Q-network (but ours will actually be shallow)**: *(Light AI programming)*. Learn how to use a neural network to make sequential decisions. In this case, the key insight is that both the table and the neural network simply represent and approximate a function, so we replace the table with a neural network. Again, we explore the impact of different parameter settings on learning performance. This lab is strongly tied to papers on the reading list, such as Vinyals et al. (2019), because it illustrates a very simple version of the type of agent those authors created
5	**Finding your own critical states**: *(AI evaluation)*. Gain insight into the functioning of an agent by selecting a test state judiciously among the vast number of states it observed, based on an approach from the reading list (Huang et al., 2018). The idea here is that to train an agent requires making a lot of action selections. Now suppose you logged all of them and want to pick which states to show to convince yourself (or someone else) that the agent is functioning well. State criticality has a few different mathematical definitions, but they always measure a difference between the best available action, according to the agent, with some other reference point (e.g., average or minimum). To clarify, driving straight on a highway with no one around is not a critical state because all choices should lead to a safe outcome. Similarly, being 6 inches away from another car headed toward you at high speed is similarly not a critical state, but for a different reason—all choices lead to an *unsafe* outcome. Critical states tend to be ones where some choices are good and some choices are bad, meaning it is important that we have trained the system to pick a good choice in that state
6	**Policy iteration**: *(AI programming)*. Understand and measure the quality of a policy iteration agent, a policy-based technique. This contrasts with the value-based techniques we worked with in earlier labs (see Fig. 4). Importantly, policy-based approaches do not track much of the historical data of action selections and observed rewards. Instead, they opt to store only the action one selects in each state. To measure the quality of a policy, one would run it and see the rewards it obtains. To optimize the policy, one randomly changes it and see if the new policy measures as better—if so, the new policy overwrites the old
7	**How to devise RQs**: *(Group activities)*. Prepare for the final project by establishing a set of possible research questions (RQs) for each student. Understand what makes a good RQ. Start thinking about how a particular study methodology might match well to a particular kind of RQ (or not). Begin finding related work for the final project based on the RQ emerging from the lab
8	**Recidivism—feature engineering**: *(AI programming)*. Work with realistic data preparation and feature engineering steps in a pipeline, as found in Dodge et al. (2019). Begin thinking about bias in ML/AI and the societal impact lack of fairness might exert

(continued)

Table 2 (continued)

#	Title: *(Focus)*. Description
9	**Recidivism—understanding explanation output**: *(AI applications)*. Work with classifier construction and explanation generation steps in the previously described pipeline. Think about which explanations are better or worse at revealing underlying issues with the model. Engage with the literature from a statistical battle between a journalistic outlet and software vendor (ProPublica and NorthPointe, specifically)
10	**Using IBM's explainability 360**: *(Explanation generation)*. Gain experience using an off-the-shelf product (IBM, 2023) to generate explanations. Learn about the efficacy of models for medical expense prediction
11	**Using LIME**: *(Explanation generation)*. Gain experience using an off-the-shelf product (Ribiero, 2020) to generate explanations. Again examine medical expense prediction, but also expand to a simple language problem using the 20 newsgroups dataset
12	**Using SHAP**: *(Explanation generation)*. Gain experience using an off-the-shelf product (Lundberg, 2018) to generate explanations. Again examine medical expense prediction and 20 newsgroups datasets. Compare and contrast the explanation strategies covered in labs 10–12
13	**Saliency maps**: *(Explanation generation)*. Gain experience using an off-the-shelf product (to be determined) to generate explanations for a simple vision dataset. Prior to this lab, the students would have only seen tabular data and language data, so here we would like to broaden to image data
14	**Capstone—explanations and NLP**: *(To be determined)*. We are still designing this lab, but the broad goal is to give the students a bit of a more open-ended language problem and see what insights they are able to generate

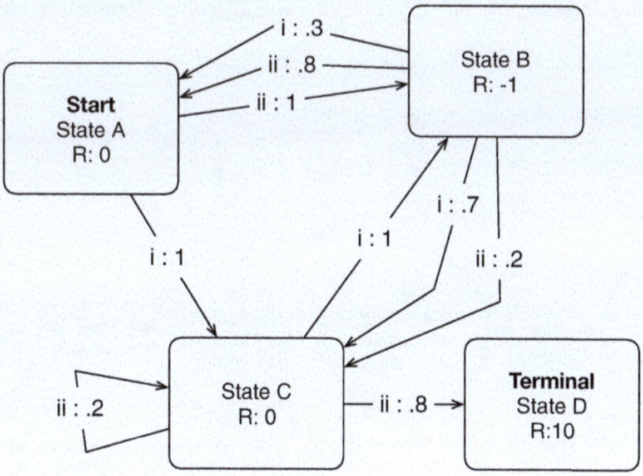

Fig. 1 A small toy Markov decision process (MDP) that has two actions. On this MDP, the right sequence of actions is to select action i in state A, which will put the agent in state C deterministically (transition probability is 1). Once in state C, it should choose action ii and either finish the trajectory (80% of the time) or land in state B. If the unlucky agent lands in state B, it should choose action i in an effort to get back to state C

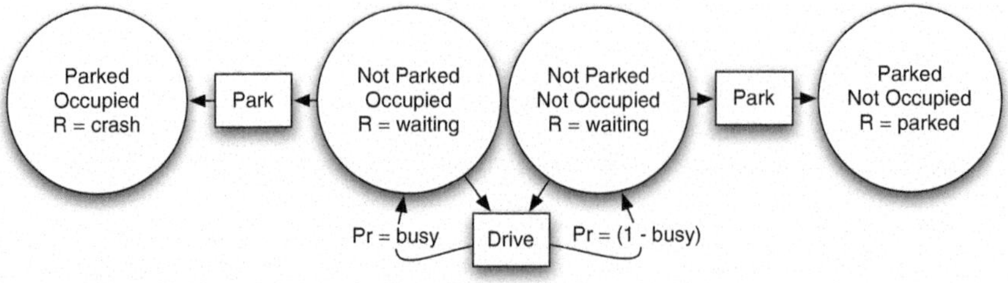

Fig. 2 MDP modeling a parking domain with a single space, parameterized by transition probabilities and reward structure

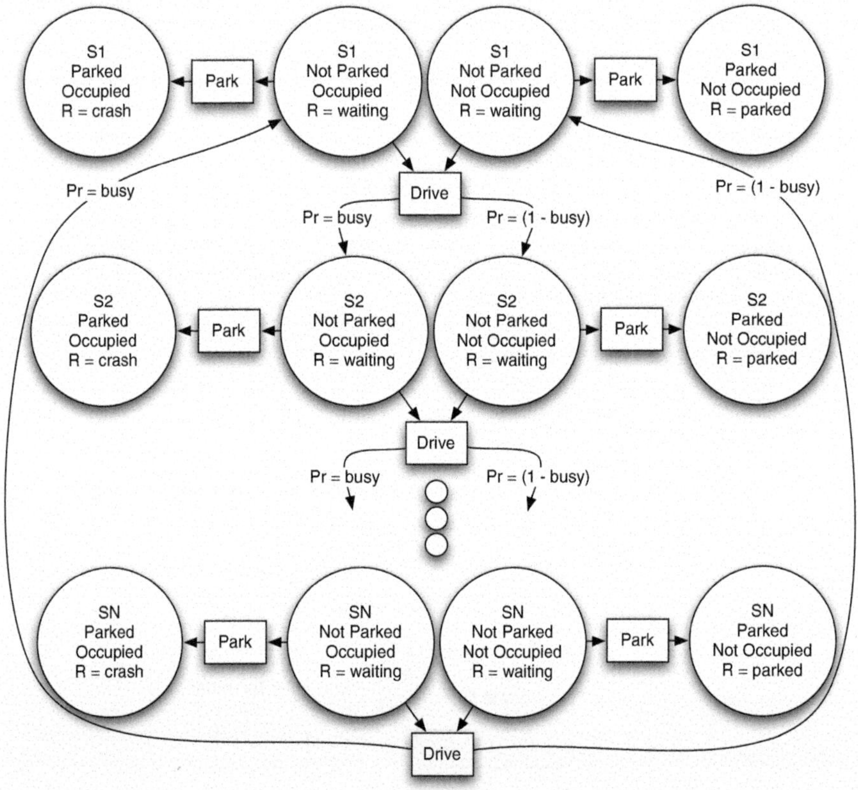

Fig. 3 An MDP formed by "gadgeting" single *spaces* (as in Fig. 2) together to form a parking *lot* for the agent to traverse as it looks for an open space

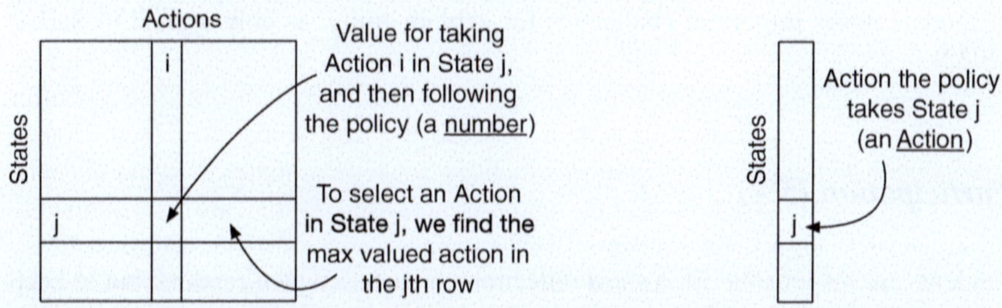

Fig. 4 **Left**: A notional illustration of a Q-learning table. Each cell (i, j) contains a *value* (a number), which the agent updates each time it observes a reward for taking Action i in state j. To select a greedy action, the agent finds the action with the max value. During training, the agent will make a mix of greedy choices and random ones in an effort to balance the "explore-exploit trade-off" as it tries to populate each cell of the table with good value estimates. **Right**: A notional illustration of a policy. During policy iteration, the agent needs to start with a policy, which can be a random one. From there, the agent will measure the value of the policy by running it. Next, the agent will change some state's action randomly and rerun the policy. If the new value is higher, the change becomes permanent; otherwise, the agent will try another change

data preparation and feature engineering (Lab 8), before moving on to the model implementation and explanation generation phases of the pipeline (Lab 9). Many of the questions in this lab focus on discussing the societal impacts of using ML to automate decisions, as well as in decision support systems (DSS) to make important life decisions like sentencing and bail. As an example of this type of question, we link the students to primary sources which illustrate how a simple interaction with a website can engender a variety of kinds of bias.[2] We then contrast that figure with the labyrinthine criminal justice flowchart[3] in order to suggest the many forms of bias that might arise from such a complex process.

The third series of four labs switches gears again and focuses on libraries that students can use off the shelf to generate explanations for the models they create. This includes: IBM's AI Explainability 360 (AIX360, Arya et al., 2019; IBM, 2023), Local Interpretable Model-Agnostic Explanations (LIME, Ribeiro, 2016; Ribiero, 2020), SHapley Additive exPlanations (SHAP, Lundberg et al., 2017; Lundberg, 2018), and saliency maps (still determining an approachable package as of this writing). In each of these labs, we mostly use tabular data for simplicity, but the students get some exposure to non-tabular data as well.

We intend the last lab to be a sort of "capstone," with focus on explanations and natural language processing. Note that language models tend to be a fairly difficult type of model to explain, in part because they are so large. The rapid growth

[2] The first image is Fig. 1 from Baeza-Yates (2018), indicating a number of different kinds of bias that can arise from a fairly simple web interaction.

[3] The second image is a depiction of the complicated interactions present in the criminal justice system (image source: https://bjs.ojp.gov/media/image/45506).

of models poses important challenges for explainability, as ably argued in Sarkar (2022).

Participation (5%)

Students are responsible for a short reflection on each assigned reading, but to keep the workload manageable (particularly if someone gets sick), we drop the lowest seven scores. We intend for the reflection to serve several purposes: checking that students did the reading, illuminating pain points to the teaching team, and providing the students with an annotated bibliography that will help them prepare their final report. The reflection covers the following questions:

1. What was it about?
2. What did it report?
3. What did you like about it?
4. What did you NOT like about it?
5. How can this paper be related to your final project topic?

Several other minor class elements are tied to participation credit, such as preparing a short introduction document to help the teaching team put a name to a face. The other elements of participation are paper presentation sign-up, attending TA office hours, and attending professor office hours.

Final Project Report (20%)

The class culminates in a final project report that takes the shape of what we call a "research paper preamble"—containing an abstract, introduction (with RQ), background, related work, methodology, and (optional) results. The idea is that if one executed the planned study, it could answer the provided RQ. In this document, the students must cite some number of papers from the class reading list, as well as provide some of their own that address their chosen topic. Students are allowed to use any AI writing assistance tools they see fit, because these tools will be available to them in the real world. The students also provide a reflection on their experience in the class in this report.

Final Project Presentation (10%)

We grade the document as an essay but also require each student to give an elevator pitch to the class. We grade these very leniently; this is mostly an exercise to improve research communication, so we try to keep it low stakes. After the presentations, the students in the audience can vote on a presentation they found "impactful" (note that impactful does not necessarily mean "best"), and we award bonus points based on that vote distribution.

Lessons for and from the First Author: Reflections

This is my second year of service, so my approaches are still evolving. This section will begin by highlighting parts of the course that we feel have been going well and then describe how it evolved through the three offerings.

What Is Going Well?

Adding a Personal Touch The participation credit for attending office hours has been going extremely well. My goal was to add a forcing function to get students to voice their questions, get help with setting up code, better understand papers, or get help with whatever else they might be struggling to accomplish. While walking from the classroom to my office together, one student recently remarked *I don't think I've ever been on the third floor before.* Note that this indicates that this student *had never attended office hours for any other class* as a student in our college.

Another aspect of the course I think is working well is the "Introduce Yourself" assignment. In all my courses, I assign this for a variety of reasons, such as helping learning names, writing more personal recommendation letters, understanding the application areas students are interested to study better, and identifying possible areas of strength or weakness.

Minimizing Programming Requirements The "Introduce Yourself" assignment continues to reveal that a decent fraction of our students are coming from computer science because they did not like programming. I have tried (and continue to try) to keep this in mind when planning curriculum and activities, as it has been a little bit of an acclimation process for me as an instructor, since my background as a student, TA, and instructor is in computer science. As such, improving students technical ability writ large in the College of IST is emerging as a core goal for me during my time at Penn State. To do so, in my courses, I try to increase the technical level while lowering the bar for entry into developing technical understanding.

Collecting Student Feedback Around week 6, we always collect feedback about how the course is going. I chose this time frame since the students have seen enough of the course format and content to inform their feedback, but not so much of it that it is too late to course correct. This takes the form of an anonymous canvas survey which asks ten questions: *what should (the professor, the assistants, and you) (start, stop, and keep) doing*? (questions 1–9), and *do you have any other comments you would like to make*? This feedback is often helpful, for example, it showed me that

I needed to give students more encouragement to talk to me about the lab—this is where the new office hours participation credit came from. The feedback is often quite honest as well, with a common theme of *I should stop procrastinating.* I think it is important to reinforce such a message both by externalizing it, but also showing how widespread this sentiment is, because nearly 100% of the *what should you stop doing?* questions address the topic of procrastination.

Presenting Papers In-class The course has a very long reading list, which will almost certainly be a hardship for busy students to actually read and understand. The in-class presentations help to make this content more accessible but gives the student a good sense of how they can learn from research papers in their future careers as the rapidly evolving fields of AI and XAI change. Most students do a good job with the one presentation they are responsible to provide and seem to listen attentively to the papers that provide highly interesting findings. The reading list continues to evolve with time and would be highly customizable if one wanted to make the course focus more on some topic areas (such as graph neural networks, which we currently do not see much of in the course).

How Has the Course Evolved over Time?

The first IST offering was a bit of a struggle as students' technical ability was really varied, and it hindered my ability to have them work with code. This is why I originally hatched the idea of having low coding requirements, but giving resources to let interested and capable programmers make deeper changes to the existing codebase, such as programming different update rules for the Q-learning assignment.

Determining the Meeting Schedule I originally designed the course to fit a MWF meeting schedule, with 45 total sessions of 50 minutes each. These iterations featured about 23 paper presentations, 6 lectures, and 12 in-class labs, to which we devoted a full class period. We spent the remaining 4 sessions on syllabus day, a bonus topic (processes for consuming explanations), student project presentations, and an open date (e.g., for a snow day, bonus topic, or longer student final project presentations). As enrollment has grown, paper presentations have been in groups, and the final project presentations have taken on the form of elevator pitches.

In the third offering, we switched over to a TuTh meeting schedule with 30 sessions for 75 minutes. To handle this, I split most class sessions, which means we spend a little less time on each paper/lab/lecture but actually have more slots. Thus, I expanded the reading list a little bit and added labs, partially because I wanted the students to see a vision application and work with a small language model more than the labs

Adding Labs with Each Offering When I was a classroom assistant as a grad student, some of my best learning experiences were from creating new assignments or overhauling old ones to improve them. As such, I try to give each assistant that works on this class with me the opportunity to do the same. The first offering did not have the DQN lab, the state criticality lab, or as many explainability strategy labs. This is also why the current lab listing is a little cagey about the content of the last two labs; we are still building the plane as we fly the plane. There are two other nice properties of designing new assignments: first, it makes cheating more difficult; second, it gives the assistant hours to report early in the semester before grading gets heavier.

Entering the semester, I had hoped to do a rather ambitious project with having the students work on MNK Games, which is a generalization of tic-tac-toe (board has size $M \times N$ and both players seek a sequence of length k to win). For this domain, I had previously implemented a lot of infrastructure for a conference paper (Dodge et al., 2022), but it is taking some time to adapt this code to create approachable assignments because it has a number of complicated library dependencies, including pyTorch, OpenGL, and wxWidgets. Perhaps, this may appear in a similar course for grad students.

Consolidating Submissions In earlier offerings, we had followed a design where the labs were simpler, intended to be done entirely in class. Then, each lab would feed into an assignment that was to serve as homework. In this framing, the labs serve as a formative evaluation. Meanwhile, the assignment components would serve as a summative evaluation. However, we found that confused some students and chose to make each lab more self-contained, but longer so that they would start in class and finish on their own.

Checking Understanding of the Reading List Content Initially, we had followed a strategy where students were responsible for posing an on-topic question for each reading. However, we found many students liked to hear the presentation before reading. As a result, questions would often be both sparse at the time of content delivery and laborious to answer after the fact. Thus, we moved a short structured reflection answering a few questions in 1–2 sentences, as described in section "Participation (5%)".

Conclusion

DS402: XAI is a course I have been working to move toward a permanent listing. This course structure seems like it would generally allow a fresh Ph.D. graduate to quickly bootstrap a course on just about any topic, since that person's dissertation and research artifacts can easily serve as instructional materials. My hope is that it can be an upper division course that offers content that can benefit students currently enrolled in the data science program, as well as the College of Information Science and Technology and the university more broadly. As an example, many HCI students should also be interested to learn these concepts. Further, I hope that this course eventually proves beneficial to the students who will enroll in the forthcoming AI degree.

Acknowledgments I'd like to acknowledge the generous and honest input from the students who enrolled in all course offerings.

References

Abdul, A., Vermeulen, J., Wang, D., Lim, B. Y., & Kankanhalli, M. (2018). Trends and trajectories for explainable, accountable and intelligible systems: An HCI research agenda. In *Proceedings of the 2018 CHI Conference on Human Factors in Computing Systems* (Vol. 582).

ACM. (2018). Acm code of ethics and professional conduct [Last Accessed: March 16 2024]. https://www.acm.org/code-of-ethics

Amershi, S., Weld, D., Vorvoreanu, M., Fourney, A., Nushi, B., Collisson, P., Suh, J., Iqbal, S., Bennett, P. N., Inkpen, K., Teevan, J., KikinGil, R., & Horvitz, E. (2019). Guidelines for human-AI interaction. In *Proceedings of the 2019 CHI Conference on Human Factors in Computing Systems* (pp. 1–13). https://doi.org/10.1145/3290605.3300233

Anderson, A., Dodge, J., Sadarangani, A., Juozapaitis, Z., Newman, E., Irvine, J., Chattopadhyay, S., Olson, M., Fern, A., & Burnett, M. (2020). Mental models of mere mortals with explanations of reinforcement learning. *ACM Transactions on Interactive Intelligent Systems, 10* (2), 1–37. https://doi.org/10.1145/3366485

Arya, V., Bellamy, R. K. E., Chen, P.-Y., Dhurandhar, A., Hind, M., Hoffman, S. C., Houde, S., Liao, Q. V., Luss, R., Mojsilović, A., Mourad, S., Pedemonte, P., Raghavendra, R., Richards, J., Sattigeri, P., Shanmugam, K., Singh, M., Varshney, K. R.,Wei, D., & Zhang, Y. (2019). One explanation does not fit all: A toolkit and taxonomy of AI explainability techniques. https://arxiv.org/abs/1909.03012

Baeza-Yates, R. (2018). Bias on the web. *Communication of ACM, 61* (6), 54–61. https://doi.org/10.1145/3209581

Binns, R., Van Kleek, M., Veale, M., Lyngs, U., Zhao, J., & Shadbolt, N. (2018). 'It's reducing a human being to a percentage': Perceptions of justice in algorithmic decisions. In *Proceedings of the 2018 CHI Conference on Human Factors in Computing Systems* (pp. 1–14). https://doi.org/10.1145/3173574.3173951

Chen, N.-C., Suh, J., Verwey, J., Ramos, G., Drucker, S., & Simard, P. (2018). Anchorviz: Facilitating classifier error discovery through interactive semantic data exploration. In *23rd International Conference on Intelligent User Interfaces* (pp. 269–280). https://doi.org/10.1145/3172944.3172950

Dodge, J., Anderson, A. A., Olson, M., Dikkala, R., & Burnett, M. (2022). How do people rank multiple mutant agents? In *27th International Conference on Intelligent User Interfaces* (pp. 191–211). https://doi.org/10.1145/3490099.3511115

Dodge, J., Liao, Q. V., Zhang, Y., Bellamy, R. K. E., & Dugan, C. (2019). Explaining models: An empirical study of how explanations impact fairness judgment. In *Proceedings of the 24th International Conference on Intelligent User Interfaces* (pp. 275–285). https://doi.org/10.1145/3301275.3302310

Eiband, M., Buschek, D., & Hussmann, H. (2021). How to support users in understanding intelligent systems? Structuring the discussion. In *26th International Conference on Intelligent User Interfaces* (pp. 120–132). https://doi.org/10.1145/3397481.3450694

Eykholt, K., Evtimov, I., Fernandes, E., Li, B., Rahmati, A., Xiao, C., Prakash, A., Kohno, T., & Song, D. X. (2018). Robust physical-world attacks on deep learning visual classification. In *2018 IEEE/CVF Conference on Computer Vision and Pattern Recognition* (pp. 1625–1634).

Goodfellow, I., & Papernot, N. (2017). The challenge of verification and testing of machine learning [Blog Post, Last Accessed: March 16, 2024].

Groce, A., Kulesza, T., Zhang, C., Shamasunder, S., Burnett, M., Wong, W., Stumpf, S., Das, S., Shinsel, A., Bice, F., & McIntosh, K. (2014). You are the only possible oracle: Effective test selection for end users of interactive machine learning systems. *IEEE Transactions on Software Engineering, 40*(3), 307–323. https://doi.org/10.1109/TSE.2013.59

Hoffman, R. R., Mueller, S. T., Klein, G., & Litman, J. (2018). Metrics for explainable AI: Challenges and prospects. *CoRR, abs/1812.04608*. http://arxiv.org/abs/1812.04608

Huang, S. H., Bhatia, K., Abbeel, P., & Dragan, A. D. (2018). Establishing appropriate trust via critical states. In *IEEE/RSJ International Conference on Intelligent Robots and Systems (IROS)* (pp. 3929–3936). https://doi.org/10.1109/IROS.2018.8593649

IBM. (2023). Software package: AI explainability 360 (v0.3.0) [Last Accessed: March 16, 2024]. https://github.com/Trusted-AI/AIX360

Jin, W., Fan, J., Gromala, D., Pasquier, P., & Hamarneh, G. (2022). Euca: The end-user-centered explainable AI framework. https://arxiv.org/abs/2102.02437

Kahng, M., Andrews, P. Y., Kalro, A., & Chau, D. H. (2017). Activis: Visual exploration of industry-scale deep neural network models. *IEEE Transactions on Visualization and Computer Graphics, 24*, 88–97. https://api.semanticscholar.org/CorpusID:3571027

Kim, M.-J., Kim, K.-J., Kim, S., & Dey, A. K. (2016). Evaluation of StarCraft artificial intelligence competition bots by experienced human players. In *ACM CHI Conference Extended Abstracts* (pp. 1915–1921).

Kulesza, T., Burnett, M., Wong, W., & Stumpf, S. (2015). Principles of explanatory debugging to personalize interactive machine learning. In *ACM International Conference on Intelligent User Interfaces* (pp. 126–137).

Lim, B., Dey, A., & Avrahami, D. (2009). Why and Why Not explanations improve the intelligibility of context-aware intelligent systems. In *2009 SIGCHI Conference on Human Factors in Computing Systems* (pp. 2119–2128).

Lundberg, S. (2018). Software package: SHAP [Last Accessed: March 16, 2024]. https://shap.readthedocs.io/en/latest/

Lundberg, S. M., & Lee, S.-I. (2017). A unified approach to interpreting model predictions. In *Proceedings of the 31st International Conference on Neural Information Processing Systems* (pp. 4768–4777).

Mitchell, M., Wu, S., Zaldivar, A., Barnes, P., Vasserman, L., Hutchinson, B., Spitzer, E., Raji, I. D., & Gebru, T. (2019). Model cards for model reporting. In *Proceedings of the Conference on Fairness, Accountability, and Transparency* (pp. 220–229). https://doi.org/10.1145/3287560.3287596

NPR Radiolab. (2020). The cataclysm sentence [Last Accessed: March 16, 2024]. https://radiolab.org/podcast/cataclysm-sentence

Penney, S., Dodge, J., Anderson, A., Hilderbrand, C., Simpson, L., & Burnett, M. (2021). The shoutcasters, the game enthusiasts, and the AI: Foraging for explanations of real-time strategy players. *ACM Transactions on Interactive Intelligent Systems, 11*(1), 1–46. https://doi.org/10.1145/3396047

Philomin, V., & Hallinan, P. (2022). Introducing AWS AI service cards [Last Accessed: March 16, 2024]. https://aws.amazon.com/blogs/machinelearning/introducing-aws-ai-service-cards-a-new-resource-to-enhance-transparency-and-advance-responsible-ai/

Poursabzi-Sangdeh, F., Goldstein, D. G., Hofman, J. M., Wortman Vaughan, J. W., & Wallach, H. (2021). Manipulating and measuring model interpretability. In *Proceedings of the 2021 CHI Conference on Human Factors in Computing Systems* (pp. 1–52). https://doi.org/10.1145/3411764.3445315

Ribeiro, M. T., Singh, S., & Guestrin, C. (2016). "Why should i trust you?": Explaining the predictions of any classifier. In *Proceedings of the 22nd ACM SIGKDD International Conference on Knowledge Discovery and Data Mining* (pp. 1135–1144). https://doi.org/10.1145/2939672.2939778

Ribiero, M. T. (2020). Software package: LIME (v0.2.0.0) [Last Accessed: March 16, 2024]. https://github.com/marcotcr/lime

Sarkar, A. (2022). Is explainable AI a race against model complexity? https://arxiv.org/abs/2205.10119

Schemmer, M., Kuehl, N., Benz, C., Bartos, A., & Satzger, G. (2023). Appropriate reliance on AI advice: Conceptualization and the effect of explanations. In *Proceedings of the 28th International Conference on Intelligent User Interfaces* (pp. 410–422). https://doi.org/10.1145/3581641.3584066

Shull, F., Singer, J., & Sjøberg, D. (2007). *Guide to advanced empirical software engineering*. London: Springer. https://books.google.com/books?id=hloWonPJyC4C

Silver, D., Hubert, T., Schrittwieser, J., Antonoglou, I., Lai, M., Guez, A., Lanctot, M., Sifre, L., Kumaran, D., Graepel, T., Lillicrap, T., Simonyan, K., & Hassabis, D. (2018). A general reinforcement learning algorithm that masters chess, shogi, and Go through self-play. *Science, 362*(6419), 1140–1144. https://doi.org/10.1126/science.aar6404

Vaccaro, K., Huang, D., Eslami, M., Sandvig, C., Hamilton, K., & Karahalios, K. (2018). The illusion of control: Placebo effects of control settings. In *Proceedings of the 2018 CHI Conference on Human Factors in Computing Systems* (pp. 1–13). https://doi.org/10.1145/3173574.3173590

Vinyals, O., Babuschkin, I., Czarnecki, W. M., Mathieu, M., Dudzik, A., Chung, J., Choi, D. H., Powell, R., Ewalds, T., Georgiev, P., Oh, J., Horgan, D., Kroiss, M., Danihelka, I., Huang, A., Sifre, L., Cai, T., Agapiou, J. P., Jaderberg, M., . . . Silver, D. (2019). Grandmaster level in StarCraft II using multi-agent reinforcement learning. *Nature, 575*(7782), 350–354. https://doi.org/10.1038/s41586-019-1724-z

Weiss, G., Goldberg, Y., & Yahav, E. (2018). Extracting automata from recurrent neural networks using queries and counterexamples. In J. Dy & A. Krause (Eds.), *Proceedings of the 35th International Conference on Machine Learning* (Vol. 80, pp. 5247–5256). PMLR. http://proceedings.mlr.press/v80/weiss18a.html

White House. (2022). Blueprint for an AI bill of rights [Last accessed: 10/13/22]. https://www.whitehouse.gov/ostp/ai-bill-of-rights/

Part III
Embedding the Classroom in the World

Traversing the Software Development Life Cycle

Reflections on Teaching Object-Oriented Programming to Third-Year IT Students

Daniel Welch

Abstract The software development life cycle provides a "template" outlining the various stages and activities involved in application development. This chapter describes and reflects on the author's efforts to utilize the life cycle—including the ordering it provides—to introduce, structure, and contextualize various topics in a programming-heavy object-oriented application development course taught to third-year information science and technology students. Through multiple redesigns, the presented approach to the course blends practical and theoretical topics in object-oriented programming with a development workflow that incorporates use of version control throughout. To this end, the chapter also includes thoughts on the utilization of tools such GitHub Classroom to help manage submission and retrieval of student work as well as lessons learned.

Keywords Teaching object oriented programming · Data structure implementation · Testing and debugging · Software Development Lifecycle (SDLC)

Introduction

Learning to program has a well-earned reputation for being notoriously difficult. In a typical introductory programming course, students not only need to grapple with the syntax of the language being used with all of its associated "incantations" (which generally must be taken as an article of faith when starting out), but more importantly: they also need to begin developing the step-by-step logical thinking and reasoning skills necessary to design and implement applications and algorithms. The need to strengthen and develop such skills does not plateau or abate after the

D. Welch (✉)
College of Information Sciences and Technology, The Pennsylvania State University, University Park, PA, USA
e-mail: dtw5246@psu.edu

J. M. Carroll (ed.), *Innovative Practices in Teaching Information Sciences and Technology*, https://doi.org/10.1007/978-3-031-61290-9_11

curriculum's introductory programming course. If anything, the subsequent course only quickens the pace with the introduction of new constructs (e.g., classes and interfaces) and techniques (e.g., inheritance, recursion, performance analysis) that vastly expand the space of possible solutions to a given technical problem.

Even when relative proficiency in a language's syntax and semantics is reached, and logical thinking becomes a fairly routine activity, students must still develop comparatively "lower-order" yet still critical language-specific tasks such as working with build scripts, organizing projects accordingly, and managing dependencies. All of this can quickly compound for the information science and technology student, who must contend with gaining relative proficiency in multiple different languages spanning different programming paradigms—oftentimes in the same semester. Indeed, where one course might use an object-oriented (OO) imperative language such as Python or Java, another might rely on a declarative language such as SQL.

This assortment of languages, paradigms, and required competencies forms the backdrop for students entering the third course in the IST programming sequence and the primary focus of this chapter: *object-oriented design and software applications*. This course, which uses Java as the primary implementation language, in some sense represents a confluence of the various object-oriented programming topics introduced in isolation in the prior two courses in the curriculum. In particular, the course treads further into the software engineering domain at large: with topics ranging from software development processes and OO design to specification and design representation languages (e.g., the Unified Modeling Language, UML). The course also places an emphasis on unit testing, version control, and the development of reusable generically typed data structures and the central role these play in developing efficient applications.

The chapter is organized as follows. In section "The Software Life Cycle and Course Structure", the software development life cycle (SDLC) is reviewed, and a high-level breakdown of course topics is outlined. This section concludes with an overview of various tools utilized in the course and a summary of student workflow. The subsequent sections "Process Modeling, Design and Specification, and Implementation"–"Final Pair Project: Nonlinear Data Structures and Implementation of a Graph Library" summarize and reflect on selected activities and assignments—including strategies used to motivate and introduce various concepts and assignments, as well as observations and lessons learned.

The Software Life Cycle and Course Structure

By the time students reach this point in the programming curriculum, basic knowledge of fundamental topics is assumed (e.g., classes, interfaces, inheritance), and the focus becomes more broadly concerned with software engineering, data structure development, and object-oriented design principles and practices.

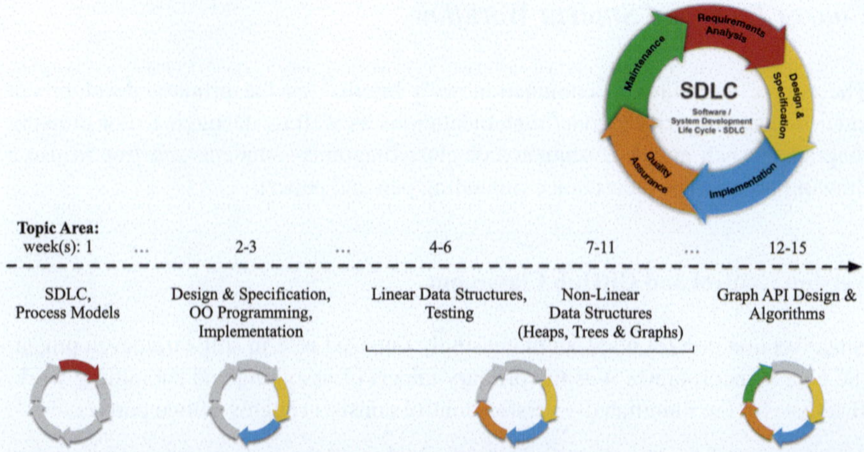

Fig. 1 The SDLC (top) and a birds-eye view of the course topics grouped by week and the SDLC topic area(s) introduced/utilized within the specific topic group

Correspondingly, one of the first and guiding topics introduced in the course is the software development life cycle (SDLC). Beyond its pedagogical role in enumerating the various high-level activities that comprise software development, the SDLC also serves as a convenient form of "connective tissue" unifying some of the more disparate-seeming topics introduced and practiced throughout the semester. A high-level outline of the semester is depicted in Fig. 1.

Under each topic area (e.g., linear data structures and testing), the corresponding section(s) of the life cycle covered and utilized are highlighted in color, while stages not yet covered are depicted in gray.

One might note from Fig. 1 that the course takes a decidedly technical approach with a heavy emphasis on the development of reusable and efficient data structures—many of which are not included in the Java collections framework (e.g., trees and graphs). And since data structures comprise the backbone of any nontrivial application, their development is used as a vehicle to exercise key topics introduced in weeks 1–4 such as specification and object-oriented design techniques, the use of version control, and the role and importance of testing (and writing tests early and often). These assignments, which typically involve implementation of an interface for some abstract data type such as a list, tree, or graph, also provide students with repeated exposure to other principles such design by contract (and how it can be used to plan and write test cases) as well as exposure to some selected design patterns.

Course Tools and Student Workflow

The course uses Java in conjunction with IntelliJ[1] as the primary development environment for writing code/implementations as well as debugging. For drawing diagrams—such as UML sequence or class diagrams—students are free to use a drawing program of their choice (including pen and paper).

Version Control and GitHub Classroom

Since version control plays an increasingly outsized role in software development, the course incorporates it as the primary means of assigning and submitting work. Briefly—for the uninitiated—version control consists roughly of two parts:

1. A hosting service such as GitHub which "hosts" the project in an online repository. The repository functions as a de facto homepage: containing everything from starter sourcecode, documentation, feature requests, bug lists, releases, and more.
2. A version control system—such as git[2]—installed locally on a developer's computer that allows them to *clone* existing repositories, *commit* any changes made (with commentary), and *push* such changes back up to the main online repository.

Git also tracks the history of changes made to a given project—allowing developers to "roll back" to a specific commit, create entirely separate branches of the codebase (e.g., to explore tentative changes or experimental features), and selectively merge any changes back to the main branch once finished.

Familiarizing students with the full range of functionality version control offers is an impractical objective for a single semester course. Rather, a middle ground is sought: to familiarize students with the core git commands and workflow by having them use it continuously throughout the semester on each assignment and lab. To make this feasible in an undergraduate setting, classroom spaces can be created that are private to the instructor and any admitted students. Instructors control who is admitted, can assign learning assistants with access privileges to student repositories, and may create multiple classroom "spaces" for different sections.

Figure 2 provides an overview of a typical classroom setup. Instructors create starter kits for each assignment (asg) or lab locally and then push each to GitHub as a "template" repository. Students "accept" each new assignment by clicking a link in Canvas. Doing so creates a private, student-specific copy of the template repository within the GitHub Classroom space for their section. Students then clone this repository to a local machine for development (typically their laptop). After

---

[1] https://www.jetbrains.com/idea/

[2] https://git-scm.com/

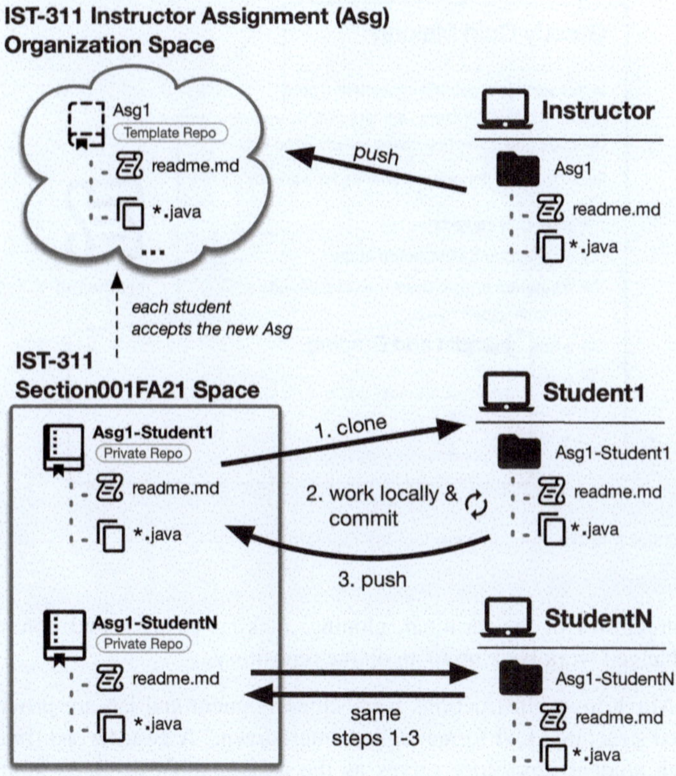

Fig. 2 GitHub Classroom setup and student workflow

working on the project, writing code, running tests, etc., each student commits their work locally (where each commit includes a short change summary) and pushes their progress back up to their remote repository.

GitHub Classroom Practices

The following are some practices I found useful when teaching and utilizing git as a key component of this course over three semesters:

Uniformity of Presentation In the first week of the course, students are required to install the git client and create a GitHub account. To enable a uniform approach when presenting git, its commands, and the general workflow—students running Windows are required to use the Unix-style bash shell that ships with git (students running macOS and Linux already have it). This practice yields an operating system independent way of learning and practicing a simple git workflow: navigating

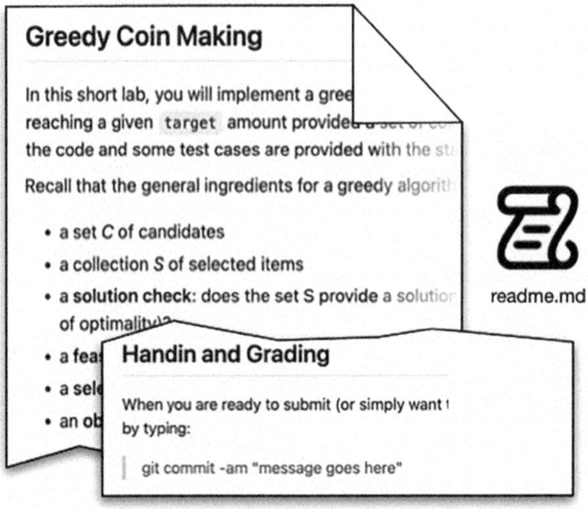

Fig. 3 Markdown instructions in a student lab repository

through directories in the terminal, cloning, making changes and commits, and pushing changed work back up to an online repository.

Utilizing Markdown Instructions for each assignment and lab are provided in a widely used structured text format called "markdown." A `readme.md` (markdown) file in each student repository serves as the assignment's de facto homepage—containing instructions, diagrams, and hints. Figure 3 shows an example.

While a seemingly trivial detail, students seemed to enjoy the unified way that this approach packages code with instructions—and how it closely mirrors the setup and presentation of real-world repositories. A more prosaic possibility is that students were simply happy to see something different from the usual Canvas UI.

Commit Thresholds To encourage routine interaction with git, a criterion in the rubric for each assignment requires students to make several commits (∼5–6 minimum per assignment) with meaningful messages and change summaries over the course of one to 2 days prior to the due date. And while students may commit and push their work as many times as they like, this practice specifically was incorporated to reduce the number of incidences of students making singular monolithic commits.

In the sections that proceed, the high-level topic areas outlined in Fig. 1 are discussed, along with summaries of selected assignments and associated practices that I've found useful in teaching the course.

Process Modeling, Design and Specification, and Implementation

After introducing the SDLC and setting up version control, the remainder of the first 3 weeks of the semester include a discussion of process models and methods for specifying and designing object-oriented systems. This initial section of the course culminates with an assignment in which students create an application from a high-level English description.

Motivating the Need for Structured Processes

A primary learning objective in the initial weeks of the course is to communicate why software is developed in the iterative manner that prevails today. To this end, the course uses Bertrand Meyer's book Agile! The Good the Bad and the Hype (Meyer, 2014) to provide a nuanced overview of the current software development landscape, as well as a detailed summary of agile methods (and some of their pitfalls).

One of the ways the course examines the importance of iteration is by looking to past exemplars of "how not to do it"—such as the waterfall model, in which each phase of SDLC is tackled one at a time and must be "completed" before work on the next phase can begin. Once this model is described, students can generally intuit the central problem: that it is inflexible and unable to cope with shifting project requirements. Meyer's text even acknowledges this directly, noting that, from a pedagogical perspective "the [waterfall] model survives mostly as a foil against which [to] argue for better approaches" (Meyer, 2014, p. 42).

If waterfall processes are a thing of the past however, a question that arises is why software development remains such a fraught process with high rates of failure (The Standish Group International, 2015), while other domains (some of which continue to employ waterfall style processes) escape this. A prime example given in class is physical construction: rarely—unless a nuclear reactor is being built (Dalton, 2022)—do physical construction projects suffer from the same cost, time, and technical overruns considered routine among software projects.

One reason for this disparity can be attributed to the non-tangible nature of software vs. the tangible, measurable, physical constraints imposed by the laws of physics on any physically based construction project.[3] Whether the constraint refers to the tensile strength of steel used to reinforce a load-bearing wall or other rigid physical tolerances, software, for better or worse, is not subject to such constraints.

[3] This is not to say that *software engineering* is not a form of *engineering* (be it chemical, physical, etc.), or that one is necessarily harder than another—but rather, that software requires certain distinctions and perhaps a more paranoid mindset.

To help communicate this distinction, I typically turn to some variation of the following analogy:

> Imaging building a house from the foundation up... in a physical setting you'd have to go pretty far out of your way and work *pretty* hard to make ringing the front doorbell flush the upstairs toilet. In software, however, unexpected interactions of this sort are easy to make happen—and in some cases are even accidental. All it takes to introduce subtle, difficult-to-debug errors is one variable accidentally referencing another in a seemingly disparate part of the application.

Such analogies I've found helpful, as many students can readily identify past occasions in which they've grappled with bugs and other counterintuitive behaviors arising from unexpected component interactions in parts of their code. Starting the course on this footing (1) motivates the need for repeatable, iterative process models that are flexible enough to adapt to change and (2) gives an early preview motivating the need and importance of testing early and often to build confidence in a system's design.

From Design and Specification to Implementation

In the initial assignment, students are given a high-level scenario—e.g., designing a point of sale system for a gift shop or a book check-in/check-out system for a library—and are tasked with deriving textual use-case scenarios. Once a set of use cases has been derived, the well-known class responsibility collaborator (CRC) method (Beck & Cunningham, 1989; Gray et al., 2003) is used to help identify which entities are needed as part of the system and the functionality each requires and provides.

Students then use their CRC cards to create UML class diagrams that outline the high-level organization of the system. And while it can be challenging to get students to arrive at a reasonable set of CRC cards, methods such as noun and verb analysis provide a good strategy for identifying the classes needed. In my experience teaching the course, once students are able to arrive at even a basic UML class diagram, most are able to translate it into a concrete object model implemented in Java.

And though Agile methods discussed in the initial portion of the course tend to de-emphasize the role of UML and other design artifacts in favor of working code and tests (Meyer, 2014, p. 60), such diagrams I've found to be invaluable—as they allow students to visualize the layout of a system and identify places where common traits and behaviors could be subsumed by a single superclass or interface.

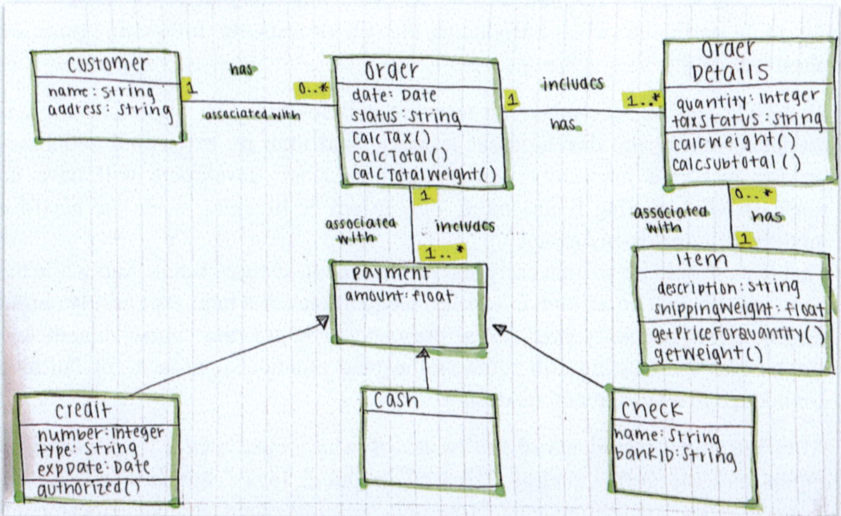

Fig. 4 A (very stylish!) UML class diagram derived from textual use cases and the CRC method

For example, in the student-drawn UML class diagram shown in Fig. 4, a payment superclass is used to encompass all valid types of payment.

Class diagrams also provide a natural place to introduce the notion of coupling and cohesion, with the goal of achieving designs that are:

1. *Loosely coupled:* meaning each class is largely independent—so changing one does not necessitate rippling changes throughout the rest of the system.
2. *Cohesive:* each class has a single coherent purpose and performs one task for readability and reuse.

While most students seem to intuitively grasp the notion of cohesion with the help of a few exemplars, learning to write loosely coupled code on the other hand is trickier. One practice I've adopted to help with this is to encourage students to use interfaces as the datatype for their variables whenever possible. This not only forces student code to rely solely on the well-defined methods and contracts exported by interfaces (which are unlikely change over time) but also reduces how reliant/tightly coupled their code is on any one particular implementation of the interface—of which there could be arbitrarily many (Bloch, 2018, p. 280).

Unit Testing and Data Structure Development

Once the first object-oriented program is written and submitted via git, the course takes a more granular focus on implementation, specification, and quality assurance—as achieved through testing (see life cycle topic areas for weeks 4–11 in Fig. 1). In particular, when introducing the role of tests, the following points are emphasized:

1. That code and its accompanying tests are perhaps the most important artifacts produced during the development process. And that, by extension, codebases lacking tests are inherently unmaintainable—since developers will have no sense of what existing functionality they might be breaking when the need for modification inevitably arises.
2. That tests should be written early, easy to run, and executed often. And while the point is made that no amount of testing can guarantee that bugs are entirely absent (indeed, tests can only ever reveal presence of bugs), the course nonetheless emphasizes that testing still remains the best practical approach for building confidence in a system and its design.

After introducing students to the basics of what constitutes a test case (a set of inputs and the corresponding expected outputs), Java's standard unit testing framework, jUnit, is introduced. From this point forward in the semester, all remaining labs and projects—most of which deal with the development of linear and nonlinear data structures (such as lists and trees, respectively)—require students to provide an accompanying jUnit test suite, implementation code, as well as a UML class diagram outlining the relationship between the classes that make up their solution.

Assignment: Implementing and Testing a Singly Linked List

The first assignment in which students are responsible for writing tests is a classic one: implementation of a singly linked list. This assignment helps establish the value of testing, as linked implementations of data structures have a well-earned reputation as challenging, oftentimes fiddly affairs full of "traps" and edge/corner cases that require students to write careful and deliberate code that benefits from extensive testing.

To help connect this section of the course on testing and data structure development to earlier topics (e.g., the role and importance of interfaces and specifications in design), an `ISTList` interface is provided as part of the starter kit for the assignment. This interface exports both the usual list methods for querying, inserting, and removing elements, in addition to more obscure (less-googleable) ones that perform some manipulation or transformation of the list (e.g., shuffling, removing every other element, interleaving elements from two lists, etc.). Each method of the interface includes a semiformal contract/specification describing its high-level behavior, as well as any exceptions that might arise when called incorrectly.

To help students get started with their own singly linked implementation of the `ISTList` interface, in class during live coding sessions, a "pinned" repository is made available to all sections in which a doubly linked implementation of the

`ISTList` interface is incrementally developed and tested.[4] After each session, any progress made on a method is pushed to a section-specific branch such that work is consistent from section to section and can be picked up exactly where it was left off when the class meets again. In-class live coding sessions like this have two key advantages.

First, it keeps students using a git-centric workflow throughout the semester: progress is committed when the last session ends, and any updates and tests added by the instructor in the interim can be pulled in via git—merging them with their own notes and progress. Some dedicated time for "syncing-up" is advised here—as many students have not yet had the occasion to merge two divergent copies of the same source file. And though the merging process has been simplified in recent years, typically learning assistants should peruse the room and assist students with lingering difficulties in merging new code.

Second, using "pinned" repositories in class to store live coding examples eliminates perennial calls to post .zip files containing code developed in class to the course canvas page—while at same time motivating the student to simply use the appropriate git command to obtain any sources they would like to consult.

The Test-Implementation-Test Cycle and the Power of Pictures

During in-class live coding sessions, I'll typically start by writing a flawed/buggy implementation of one method and immediately start to write a test for it. This first test rarely passes—necessitating a return to the code to diagnose why it failed. This involves a combination of using the IDE's debugger and/or illustrative pictures of the sort shown in Fig. 5.

Having students draw pictures is a useful way to slow down and encourage them to think incrementally about the code being developed. For example, having students draw the state of a sample list on paper, write a line or two of code that changes the list in some way, and then show the effect these lines have on the state of the list can help solidify the student's mental model of their program while also revealing potentially long-held misperceptions—e.g., on the role of references or the semantics of variable assignment (Mehta & Sahni, 2017, ch. 6). This process, by extension, also helps dispel the mistaken notion that loops or seemingly complex logic is always necessary to make large/meaningful adjustments to the data structure at hand.

As students continue to develop their code by drawing illustrative pictures and debugging, tests are continually added to exercise and build confidence in the correctness of the code begin written. Testing plans are created based on the contract of the method in question. For example, if the student is tasked with implementing

[4] While doubly linked lists are indeed similar to the singly linked version each student must complete, they are sufficiently different—and moreover, students are still required to implement and test the more obscure "coding interview" style list methods themselves.

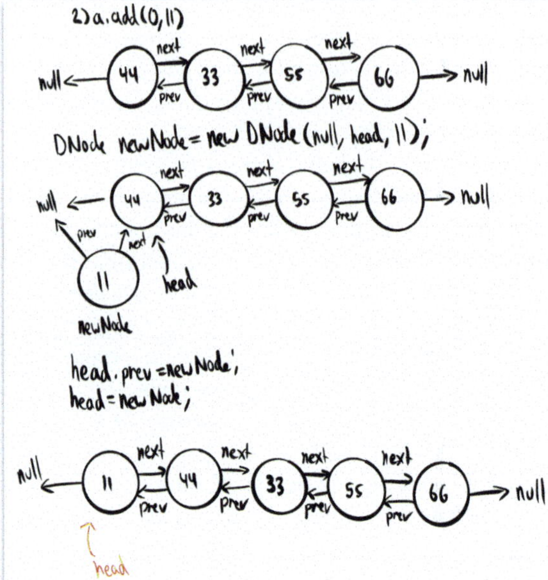

Fig. 5 A student drawing outlining the steps for adding 11 to the front of a doubly linked list a, where initially $a = [44, 33, 55, 66]$

a method that inserts an item into a list at some position i, multiple tests should be written exercising: insertion into the empty list, insertion at the front, middle, and end of a nonempty list, as well as jUnit "failure" tests that intentionally check to ensure that the correct exception is thrown by the method in the event that someone calls it incorrectly—as specified in the interface contract of the method.[5]

Student Challenges in Writing Tests

Students, especially with this first linked list assignment, often struggle when the time comes to write and run tests. And while part of this boils down to simple inexperience (writing jUnit tests can initially be a somewhat foreign experience), it can also be attributed to the dual role students must play as both implementor and tester. Indeed, the act of producing code and devising logic is an inherently creative activity, while testing is an inherently destructive one that actively requires students to attempt to break and/or reveal flaws in their own code (where the "reward" for doing so only signifies that more work is required). As a result, students often must adopt a type of split personality when writing tests: with one side deviously seeking to undermine hard-fought gains and the other continuing to strategize—eventually honing in on a potential solution to the method or problem at hand.

[5] E.g., a method contract might stipulate that some integer i passed in is within bounds, detailing the exception that should be triggered if provided with some i out of bounds.

However, when an assignment is "checked out" on the last day before it's due (something GitHub Classroom shows instructors), student inclination to skimp on meaningful and thorough tests is only heightened. And while various incentives can be given, overcoming this necessarily lies in the mindset of the student: they must have some investment in the code and its quality. Further, through lecture and practice, the reasonably invested student should not be thinking of testing as some discrete, monolithic task to be started only after all the code is written—but rather, as an iterative, continuous process that occurs regularly throughout the project as new functionality is added.

Final Pair Project: Nonlinear Data Structures and Implementation of a Graph Library

After the linked list assignment, nonlinear (hierarchical) data structures are introduced, such as trees and heaps. Assignments involving these, while similar in structure and presentation to the linked list assignment, layer in additional topics ranging from object-oriented design patterns that enhance reusability (such as providing listeners/visitors that can mechanize traversals over trees) to performance benefits that treelike data structures provide over linear structures like lists.

Capping off the transition to nonlinear data structures is a final pair assignment that tasks students with implementing a graph data structure presented and modeled in a way that it might hypothetically appear if included as a standard component in the Java collections framework. Figure 6 shows a rough UML class diagram that is provided in the starter kit for the assignment that lays out the general structure of the graph library that students must build. Note that all public-facing operations for building and manipulating graph structures are exported through the IGraph interface for clients/users of the structure.

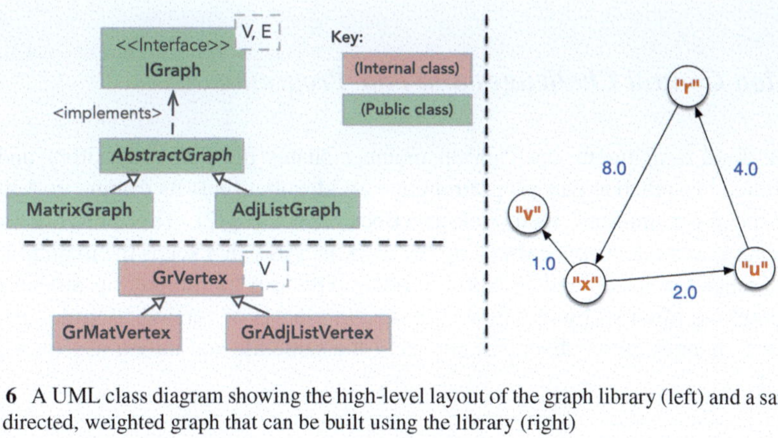

Fig. 6 A UML class diagram showing the high-level layout of the graph library (left) and a sample of a directed, weighted graph that can be built using the library (right)

The library allows users to construct both directed and undirected graphs with weighted edges. Figure 6 (right) shows an example of an edge-weighted, directed graph with four "nodes" that contains a cycle (e.g., $x \rightarrow u \rightarrow r \rightarrow x$). Graphs (potentially involving cycles) like the one shown can be easily used to model networks such as city maps, wiring diagrams, and structures from many other domains. And while students might already be familiar with the mathematical underpinnings of a graph through a prior course in discrete mathematics, this assignment is typically their first exposure to representing them in code using object-oriented design techniques.

In introducing students to the assignment, an implementation of the `IGraph` interface represented using an adjacency matrix (used to store the edge and node data for the graph itself) is developed in class through live coding sessions and activities. Students then pair up in teams to two to develop a second (adjacency list based) implementation of the `IGraph` interface (see Fig. 6, left). The project consists of two parts: (1) development and testing of the `AdjListGraph` and (2) using the library to implement an application in which graphs play a central role in the solution.

The assignment's second half in general is a good place to layer in more advanced topics in algorithms and algorithm design such as topological sorting (useful in modeling networks of dependencies—e.g., prerequisites in a curriculum), different types of searching methods (e.g., depth-first and breadth-first search), as well as other classical optimization problems.

For example, one semester had pairs implement an application that runs the trivia game: "Six Degrees of Kevin Bacon." The problem is represented using a graph in which actors are nodes and one actor is linked to another via an edge if both appeared in the same movie. The goal of the game, given an arbitrary starting actor, is to find the shortest route through the graph to the actor Kevin Bacon. To solve this, students must add functionality to build a suitable graph from a medium-sized IMDb[6] file containing several thousand movies and their actors. Teams then must implement and run breadth-first search on resulting graph to find the shortest route.

Version Control Challenges and Pair Programming

In this final assignment, each group member shares the same repository and is encouraged to practice pair programming—an idea that finds its origins in Agile's "extreme programming" methodology (Beck, 2005, p. 42). This presents some challenges, as students must divvy up the work in a way that keeps the main branch of the repository clean and in working order. This can be difficult, as students at this point are often still unfamiliar with certain aspects of version control, such as merging changes in two divergent copies of the same file and initiating requests to

[6] https://www.imdb.com/

merge completed work back to the "main" branch. Learning assistants and other instructor provided documentation is critical to support (and reassure) students that their work will not be lost—assuming regular commits have been made along the way.

Over the semesters in which I taught this course, some common pair programming strategies tended to emerge. For example, anticipating eventual file-merging challenges, certain groups would work to ensure the same file is not being edited at once, e.g., with one member electing to take the lead in implementing the core library functionality while the other writes tests—reuniting to implement the second half of the assignment in front of the same laptop.

Other groups would instead divvy up the methods within the same file and would be responsible for writing the accompanying tests for their assigned methods. One observed danger with this approach is that it tends to produce tests that narrowly focus on exercising single methods, as opposed to combinations thereof, thus sacrificing coverage. Oversights like these quickly become apparent however in the second half of the assignment—which require larger more complicated graphs that tend to reveal lingering bugs in the graph library implementation.

Concluding Thoughts

While I enjoyed teaching this course, there are a number of changes I would incorporate in future iterations. First, after the course's initial section on object-oriented design and specification, I would add a week singularly focused on reviewing performance analysis. This is crucial for the weeks that follow in order to fully understand performance distinctions between nonlinear and linear data structures.

Second, I would attempt to add more of an emphasis on object-oriented design patterns. Most students have already seen and practiced larger "architectural patterns" in the context of prior courses—which focus primarily on GUI-based application development using the Model View Controller architectural pattern. This upper-level course however is a natural place to cover more granular design patterns and the benefits they provide.

Third, being only a one semester course, the maintenance section of the SDLC invariably tends to get short shrift. One potential way of incorporating it more fully throughout would be to provide students with the opportunity to document any issues that arise via GitHub's issue tracking page based on feedback from their assignments (or their own observations). Extra credit could even be provided for addressing, writing tests, and closing such issues. Alternatively, issue tracking could be incorporated as a requirement on the longer-running final pair assignment.

References

Beck, K. (2005). *Extreme programming explained: Embrace change* (2nd ed.). Addison-Wesley professional.

Beck, K., & Cunningham, W. (1989). A laboratory for teaching object oriented thinking. *ACM Sigplan Notices, 24*(10), 1–6.

Bloch, J. (2018). *Effective Java* (3rd ed.). Boston, MA: Addison-Wesley.

Dalton, M. (2022). Nuclear power is poised for a comeback: The problem is building the reactors. *The Wall Street Journal*. Retrieved 07-16-2022, from https://www.wsj.com/articles/nuclearpower-climate-change-russia-energy-11655995024

Gray, K. A., Guzdial, M., & Rugaber, S. (2003). Extending CRC cards into a complete design process. In V. Dagdilelis, M. Satratzemi, D. Finkel, R. D. Boyle, & G. Evangelidis (Eds.), *SIGCSE Conference on Innovation and Technology in Computer Science Education* (p. 226). ACM. Retrieved from https://doi.org/10.1145/961511.961582

Mehta, D. P., &Sahni, S. (2018). *Handbook of data structures and applications* (2nd ed.). Chapman & Hall/CRC. Retrieved from https://doi.org/10.1201/9781315119335

Meyer, B. (2014). *Agile!—The Good, the Hype and the Ugly*. Springer. Retrieved from https://doi.org/10.1007/978-3-319-05155-0

The Standish Group International. (2015). *The CHAOS report 2011–2015*. Retrieved from https://www.standishgroup.com/sample_research_files/CHAOSReport2015-Final.pdf

Blurring the Boundaries of Teaching Modes to Improve the Experience of Professional Learners

Guoray Cai

Abstract Professionals have benefited from Penn State world campus courses and degree programs to advance their careers. Traditionally, students in professional master's degrees mainly used asynchronous online and web-based instructions as their primary mode of learning. The last few years of fighting with COVID in university instructions has disrupted the dichotomy of resident vs. online learning and catalyzed multiple teaching modes (resident, remote synchronous, asynchronous, and hybrid). It is important to understand how these teaching modes impact the learning of resident students. In this paper, I argue that the creative use of hybrid modes has opened up new possibilities for professional students, but this practice must be carefully designed based on a deep understanding of the learning experience of professional students. I will support this argument with my own experiments on how these teaching modes could be seamlessly combined and some reflections on how delivery modes encourage or impede effective learning of professional students.

Keywords Hybrid teaching modes · Professional learners · Instructional design

Introduction

The accelerated growth of information technology industries in recent years has created new career opportunities and new demands for continued education and training at various stages of professional development. Penn State is among the first in the nation to offer master's degrees in professional studies to provide graduate education. Penn State's distance education program, World Campus, currently offers 56 professional master's degrees, and about half of the enrolled students are in information sciences and technology-related programs (AI, data analytics, cybersecurity,

G. Cai (✉)
College of Information Sciences and Technology, Penn State University, University Park, PA, USA
e-mail: gxc26@psu.edu

J. M. Carroll (ed.), *Innovative Practices in Teaching Information Sciences and Technology*, https://doi.org/10.1007/978-3-031-61290-9_12

information science, geospatial analysis in homeland security, etc.). They emphasize practical skills designed for current and aspiring professionals, including post-bachelor and postgraduate students. Traditionally, students in professional master's degrees used asynchronous online and web-based instruction as their main mode of learning. Such teaching modes fit well with the lifestyle of professionals who need flexibility and accessibility to organize their learning activities together with work and life. However, the effectiveness of such teaching modes varies significantly between different disciplines and practical fields.

I have been teaching visual data analytics for security and risk professionals via Penn State World Campus since 2012. In the meantime, I also teach large resident classes using new media technologies. This combined experience triggered much of my curiosity in why certain teaching modes do or do not work in different parts of a course or curriculum. Like most other universities, Penn State has been running resident instructions and online instructions as separate entities with different support structures and expectations. My experiences on both online and resident teaching have also stayed separate until the recent pandemic outbreaks. During the three years of COVID disasters, universities were forced to move resident classes into various remote learning modes (Watermeyer et al., 2021). Such a move was a cultural shock for most instructors who had no experiences in teaching online courses. Nevertheless, universities had no choice but making a swift change. In Penn State and elsewhere, instructors may choose to mix and match different teaching modes (resident, remote-synchronous, remote-asynchronous) to fit different teaching methods. This opened up the many opportunities for experimenting with a wide range of remote education delivery practices. Such practices have triggered a new wave of research on remote learning and delivery modes (Lockee et al., 2022). While there were many innovations, the community was quick to realize that it is not as simple as changing delivery modes. There are a suite of variables that influence the choices of delivery modes, instruction methods, and materials, and such choices are likely to be a complex and discipline specific.

Given the above context, this chapter explores ways to design and deliver information technology graduate education remotely for professional students. I will dive into the unique aspects of teaching and learning in information technology-related courses and the special characteristics of professional learners. Findings are directly derived from the author's experience in teaching multiple resident courses remotely during the COVID pandemic years and simultaneously in teaching online professional master classes through the Penn State World Campus. I will articulate my findings along several dimensions of consideration that influenced my choices of teaching methods and delivery modes and subsequently use them to justify the experiments that I have made over the last few years. I report on a few lessons learned and suggest additional questions to investigate.

Teaching Modes: Instructional Methods, Modes, and Materials

Following the work of Head (2002), we define **teaching modes** of an education environment as the joint effect of three interdependent and interacting ingredients: (1) the instructional method, (2) the delivery mode, and (3) the media attributes of the teaching materials. I will use these three dimensions to situate subsequent discussion on teaching modes.

Instructional Methods Instructional methods refer to strategies or techniques used to facilitate the intended learning outcomes. Common types of instructional methods include *instructor-centered methods* (in the form of lectures, questioning, and demonstration), *interactive methods* (such as class discussions or collaborative activities), and *programmed instruction and modularized instruction*. Each of these instructional types facilitates certain way of learning (Head, 2002). For example, programmed instruction and modularized instruction are examples of individualized learning methods-techniques that address individual learning styles by allowing students to engage in instruction at their own pace (Table 1).

Delivery Modes Effective learning in higher education is largely dependent on the choice of appropriate delivery mode (Imran et al., 2023). *The face-to-face* (F2F) mode has been traditionally considered as the most reliable one where the content is provided by the instructor in person in regular classrooms. Online/remote teaching uses technology to deliver instructional contents and facilitate learning activities. Despite the educational merits of online learning models, they have faced criticism (Daniel, 2020). It lacked the in-person touch, social learning, and individual attention of traditional teaching methods. Some disciplines or subjects were even found to be difficult to teach in an online mode (Piergiovanni, 2018). Broadly speaking, the choice of delivery modes does have impact on learning experiences, but findings about the relationships between learning and delivery modes are complex and mixed (Imran et al., 2023). To meet the requirements of almost all the disciplines, various forms of blended teaching mode (such as blending online with in-person learning) have been used to create the best settings for improving students engagement and learning outcomes (Alducin-Ochoa and Vázquez-Martínez, 2016).

Table 1 Instructional methods and their supported learning styles (Head, 2002)

Instructional methods	Learning styles
Instructor-centered method	Lectures, questioning, and demonstration
Interactive methods	Class discussions or collaborative activities
Modular instruction	Individualized instruction that allows students to use a self-contained package of learning activities
Programmed instruction	Individual learning styles—allowing students to engage in instruction at their own pace

Blended instruction is defined as the appropriate mix face-to-face instructional methods and various remote learning technologies to support planned learning and foster subsequent learning outcomes. Blended learning modes may take many forms (Trauth and Booth, 2014). One way of blending online and face-to-face learning is to have use online learning materials as the "backbone" while teachers use classroom meetings to guide and coach students on assignments and projects. Another method of blending is to supplement resident instructions with online teaching in the "a la carte" model (Piergiovanni, 2018). There is a wide design space for blended learning, and it is essential to adapt it to fit the emerging scenario by combining the strengths of multiple delivery modes with those of technology to enhance the learning experience.

Teaching Materials: Media and Structure Instructional materials include lectures, readings, textbooks, workbooks, practice sets, labs, exams, multimedia components, and other resources in a course. These materials are chosen to support three rounds of learning (Imran et al., 2023): (1) building knowledge and literacy, (2) acquiring basic abilities, and (3) applying the above in problem-solving. The last stage can be further decomposed into replicating, transferring, and applying to new problems.

Instructional materials must be organized in a way that fits the teaching strategies of the instructors and the learning strategies of the students. For example, Rosson (2014) argued that teaching interactive information systems courses using *Learning by Design* should be matched with instructional materials that are organized to support the critical analysis of multiple aspects of information systems, such as what people do with the technology, what information needed by the intended activities, and the technical elements of information system solutions.

The choices of teaching modes should be a conscious design made by the instructors based on a set of considerations (Head, 2002). For example, to support self-paced learning in online instructions, instructional materials are typically organized as *modules* where everything about a particular thematic activity is packaged together, with searching and linking capabilities to support associative learning. When delivery modes changed, there should be matching changes in instructional methods and teaching materials.

Considerations in Choosing Teaching Modes

The choice of instructional methods, modes, and media is based on three considerations. One is the nature of the subject, which may require certain types of methods. For example, there is considerable work to document the different instructional needs in business, medicine, or engineering (Imran et al., 2023). The other consideration is to match teaching modes with the learning strategies of the students (Neroni et al., 2019). Knowing which learning strategies used by students allows instructors to choose effective supportive techniques in their curriculum. Learning strategies can be classified into three types: cognitive strategies, metacognitive strategies, and resource management strategies. Metacognitive self-regulation involves planning and monitoring their own learning. The third consideration is the characteristics of the learners and the learning outcomes they sought. In the context of this chapter, we are particularly interested in professional learners.

Characteristics of Professional Learners

Professional learners are a special kind of adult learners who seek job-related problem-solving skills as a critical learning outcome (Gardner et al., 2022). Compared with other kinds of students, professional students are much more active and experienced learners (Knowles, 1972, 1980). In particular, professional learners are unique in six dimensions:

1. **Readiness to learn**. Professional learners are more aware of why a topic is important and are more motivated to learn.
2. **Highly responsible**. Professional learners take great responsibility for their own learning.
3. **Experienced learners**. Professional learners have years of experience, which can become an important resource for learning. Their experience is an important motivator (Kawalilak and Groen, 2020).
4. **Orientation to learning**. Professional learners are more problem centered than subject centered in learning.
5. **Motivation**. Professional learners are self-motivated to keep their career growing and developing.

Meanwhile, adult learners typically face unique challenges (Gardner et al., 2022). Some of them have to juggle family responsibilities (parenting or serving as a caregiver for a partner or parent), reside off campus, and navigate financial independence (Selwyn, 2011). Adult learners were also more likely to face significant time constraints because they often work full- or part-time to fund their education.

For the above reasons, adult learners favor online programs due to the perceptions of online learning as an autonomous, self-paced environment, an instructor-led learning environment, and/or an effective multimedia environment. They need educational programs that offer the "flexibility" and "adaptability" to fit well with the lives of individual learners.

Opportunities for Innovating Distance Education

Through a wide range of experiments to move into remote learning in the past, we have learned some important lessons. These lessons inform us about the needed attention in innovating distance education in general and Penn State World Campus programs in particular:

1. *Distance education does not work equally well for every discipline.* Although many studies consistently advocated face-to-face education, the online mode of education was found to be much more difficult in such disciplines as health and architecture (Kumari et al., 2021; Masdéu and Fuses, 2017). In particular, learning a technology-intensive subject using online delivery mode can be more difficult than taking a traditional classroom course (Lee and Nuatomue, 2021).
2. *Residential education has been permanently transformed by different degree of leveraging remote learning.* To meet the requirements of almost all the disciplines, various forms of blended teaching mode (such as blending online with in-person learning) have been used to create the best settings for improving students' engagement and learning outcomes (Alducin-Ochoa and Vázquez-Martínez, 2016).
3. *What we learned from teaching resident classes remotely has not been transferred to potential changes in remote learning of professional students.* Remote learning is now taking a variety of forms that were not available before the pandemic (Imran et al., 2023). Technologies are much richer in supporting synchronous and asynchronous remote teaching, virtual labs, virtual teams, flexible exams, and customized assignments. These new experiences need to be reflected on from the perspectives of professional learners who are pursuing online degrees.

Teaching Visual Analytics for Security and Risk Professionals

Given the broad trends in enabling effective learning for professional learners, I present a case study to highlight the unique challenges and opportunities in teaching professional master classes through Penn State World Campus.

In 2012, I developed IST 868 (applied visual analytics) to be taught online via Penn State World Campus. The course was design for students enrolled in Masters of Professional Studies in the fields of intelligence analysts, business decision-making, and security investigation. Analysts are faced with a continuously growing set of data originated from a wide range of sources and in a wide variety of formats. Such data need to be analyzed in order to understand a situation and to take actions promptly. Visual analytics (VA) offers an effective solution for making sense of massive datasets by augmenting human cognitive abilities with information visualization and interaction technologies. It helps analysts in detecting the expected and discover the unexpected, providing timely, defensible, and in communicating findings and assessment effectively for action. IST 868 brings practical data analytic skills, such as data preparation, visualization, synthesis, pattern seeking, interactive exploration, presentation, claim-making, and insight generation, to practical applications in business, cybersecurity, and intelligence analysis.

Up till the year 2019, IST 868 was offered using web-based learning technology. Web-based learning (WBL) (Cook, 2007) uses asynchronous remote delivery mode, coupled with tutorials, online discussion groups, and virtual labs or problems. While students enjoyed the flexibility of participation offered by WBL, the result of their learning was mixed. This was mainly due to the mismatch among learning objectives and strategies and the instructional methods and materials of web-based instructions. Next, I will elaborate on the instructional materials and methods used in IST 868 course and discuss the considerations for choosing proper delivery modes.

Three Layers of Learning in Visual Analytics

Visual analytics in risk and security domains is a type of human-centric data analysis activities (Cai, 2014) (see Fig. 1). Figure 2 is an example of the interactive visual workbench that students will learn to develop in order to support threat detection and prediction activities illustrated in Fig. 1. Students are expected to custom-build such analytic tools based on the need of threat investigation.

Given such learning objectives, students must learn not only the use of sophisticated tools and processes to analyze datasets using visual representations of the data but also the analytical reasoning skills for deriving actionable insight from data in graphs, charts, and maps. Students would not be able to accomplish above without adequate literacy on the basic concepts and theories behind visual analytic methods and tools. Therefore, IST 868 was designed to facilitate learning in three layers:

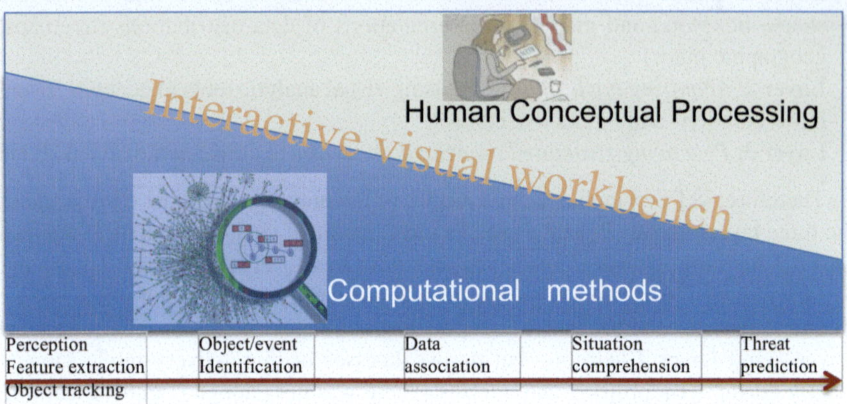

Fig. 1 Visual analytics for threat detection

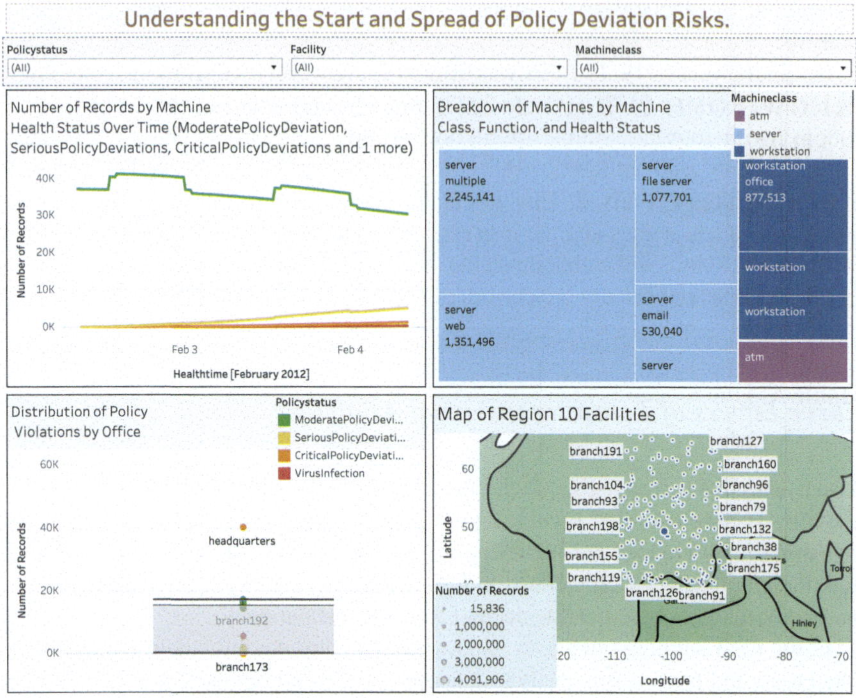

Fig. 2 An interactive visual dashboard for investigating network threats. It includes four charts (line chart, treemap, boxplot, and map views) with linking and brushing interactivity (Koytek et al., 2018)

- **Layer 1**: *Building knowledge and literacy* on statistical charts (bar charts, line charts, boxplots) and graphical representations of data distributions (treemaps, geographic maps)
- **Layer 2**: *Acquiring basic abilities* in using visual analytic tools to generate visual representations and interactive controls
- **Layer 3**: *Practicing structured analytical reasoning* to derive actionable insights

Therefore, IST 868 includes three rounds of learning activities that correspond to the three layers above. *Round 1* uses lectures, readings, and discussions to achieve *Layer 1* learning. *Round 2* uses labs and practices to achieve *Layer 2* learning. Finally, *Round 3* uses two analytic projects to achieve *Layer 3* learning.

Our experiences show that web-based instructions worked reasonably well for Layer 1 and Layer 2 learning, but the weaknesses of this delivery mode showed clearly during term projects and quizzes. There was not enough cognitive learning happening in the level of reasoning about patterns and associations in the data. Students felt lost when receiving project assignments (Layer 3). I addressed these problems by intensive coaching via office hours and emails on one hand and by adding alternative project assignments to reduce cognitive learning difficulties.

During 2019–2022, rapid transformation of resident teaching happened across the whole university (Penn State) and within the College of Information Sciences and Technology. The swift transition was made possible by leveraging advanced instructional design technologies. The creation of high-fidelity and interactive contents for remote delivery became an expected skills for all instructors. The use of such improved instructions and media of learning materials in resident classes was extremely successful. I found that students enjoyed the flexibility and accessibility of multimedia instructional materials, and they reported positive experiences in the way I blended online and F2F instructions.

Inspired by the lessons learned in resident teaching, I made a few attempts to introduce new ingredients to the traditional web-based instructional practices in teaching IST 868. Next, I report on such experiments and reflect on why the changes bring positive or negative responses.

Experimenting Blended and Blurred Teaching Modes

In the past two years, I have added a number of enhancements to teaching IST 868 (applied visual analytics). While the dominant delivery mode remains web-based learning, my goal was to explore the benefit of mixing other modes of teaching into the course. These new ingredients are targeted at overcoming the shortcomings of the traditional web-based learning.

Experiments on New Instructional Methods

E1 *Mixing lectures and labs with self-assessment activities*. This is inspired by the observation that classroom instructions allow instructors to assess students' learning by observing their activities and asking gauging questions spontaneously. However, online, web-based learning typically relies on learners to do self-assessment, due to lack of instructor involvement at the moments of taking in new learning materials. This is different from typical assessment through exams and quizzes. In response to this need, I redesigned my lectures and labs to ensure that there are probing questions or "on-your-own" activities immediately after a new concept or technique is introduced.

E2 *Interleaving demonstration, coaching, and self-guided exploration activities*. When teaching analytic skills supported by a tool (such as Tableau), professional learners have a strong sense of transferring what they learned into something meaning to their current jobs. Based on suggestions from some students, I added a component of instruction to explore new datasets or problems related to their current job duties. This is done through extra credits. I offered special zoom sessions to coach students in applying visual analytics to their job responsibilities. Although I do not collect their results related to their business assignments (due to confidentiality), I did receive positive feedback on such enhancement. One student commented: "I was given a work assignment two weeks ago to aggregate and report some data. The project at work very closely parallels the term project here and I feel prepared to take it on from what I learned in the class."

E3 *Adding midterm exams.* Most web-based courses offered on World Campus do not have exams administered by the instructors in the format similar to resident instructions. This is due to difficulties in finding a common time for professional students. However, I observed that missing exams as part of the formal assessment diminishes the effectiveness of learning data analytics. As an experiment, I added one midterm exam to IST 868. The exam questions were designed to test students' transfer learning by using new datasets and new problem scenarios. They were administered in a common time frame, with real-time access to instructor's help. The exam was open book and open notes, and students must finish the exam in a limited time frame (approximately 90 minutes). Students' responses to this addition were mixed. While some students (about 10%) objected to the idea of formal exams, most students felt that the exams did help in consolidating the skills and knowledge they learned. I observed that having such exams before students engage large term projects did improve their ability to generate in-depth analytic reports.

Experiments on New Delivery Modes

E4 *Adding random "synchronous remote" sessions.* The hardest part of teaching visual analytics is that students were unable to gain adequate learning in "Layer 3" (see section "Three Layers of Learning in Visual Analytics"). In response, I added a few evening class meetings using *remote synchronous mode* (via Zoom). I offered such meetings once every 3 weeks. I used such meetings to do cognitive training and development. Example questions from the students are how to make choices of visualization types based on the nature of the data and the questions asked; how to arrange complementary visualizations on a dashboard; and how to relate findings from more than one visualization. These types of learning require higher-level metacognitive thinking, and they are hard to learn without being guided and challenged. Even though I made participation optional for such meetings, most students attended and actively engaged. Student appreciated the meetings as a chance to reflect and connect what I have learned. One student reported that "I feel I learned more in one hour than than the hours I spent over weeks."

Experiments on New Media of Instructional Materials

E5 *Adding streaming media for video-based learning.* In earlier years of offering this course, most course materials were in text and voice-based instruction media, coupled with some view graphs and animated slides. Such media forms were hugely inadequate for learning sophisticated visual analytic methods and tools. Later, I adopted video lectures recorded from classroom teaching. The media and structure of learning materials used in IST 868 became better aligned with with the instructional methods and learning strategies of professional students.

Although enhancements on instructional materials were welcome by all students, they also experienced a few issues. *First,* the video recordings have quality issues (poor focus, resolution, and noisy). *Second*, the classroom recordings tend to have gaps of time due to interruptions. *Third*, the lecture media were too long and were hard to use. With the help of professional instructional designers, I recreated the video media for lectures and lab materials following a number of principles: (1) compact and concise; (2) modular structure; (3) breaking long videos into short lecture nuggets with searchable index and hyperlinks to associated materials; and (4) incorporating probing and assessment questions in the video. The resulting video media received highly positive responses from students. It costed months of careful planning from the instructor and the semester-long work of a team of media producers, but the outcome is well worth the investment.

Conclusions

Teaching technology-intensive subjects (such as visual analytics) for professional learners in distance education requires consideration of the unique nature of the subjects, the motivations of professional learners, and the learning strategies of experienced learners. In this chapter, I described a few ongoing experiments to innovate traditional web-based teaching modes with new ingredients of instructional methods and media that match better with the characteristics of professional learners (described in section "Characteristics of Professional Learners"). The true effects of such changes are still far from clear, but informal observations showed good promise. More experimentation is needed to search of better combinations of instructional methods, delivery modes, and media use, but the right recipe is likely to be different among educational fields.

References

Alducin-Ochoa, J. M., & Vázquez-Martínez, A. I. (2016). Hybrid learning: An effective resource in university education? *International Education Studies, 9*(8), 1.

Cai, G., Gross, G., Llinas, J., & Hall, D. (2014). A visual analytic framework for data fusion in investigative intelligence. In *Proceedings of SPIE Conference on Visualization and Data Analysis*.

Cook, D. A. (2007). Web-based learning: Pros, cons and controversies. *Clinical Medicine, 7*(1), 37–42.

Daniel, S. J. (2020). Education and the COVID-19 pandemic. *Prospects, 49*(1–2), 91–96.
Gardner, A. C., Maietta, H. N., Gardner, P. D., & Perkins, N. (2022). Online postsecondary adult learners: An analysis of adult learner characteristics and online course taking preferences. *American Journal of Distance Education, 36*(3), 176–192.
Head, J. T. (2002). Method, media, and mode: Clarifying the discussion of distance education effectiveness. *The Quarterly Review of Distance Education, 3*(2), 261–268.
Imran, R., Fatima, A., Elbayoumi Salem, I., & Allil, K. (2023). Teaching and learning delivery modes in higher education: Looking back to move forward post-COVID-19 era. *International Journal of Management Education, 21*(2), 100805.
Kawalilak, C., & Groen, J. (2020). Adult learning. In *The handbook of adult and continuing education*, T. S. Rocco,, M. C. Smith, R. C. Mizzi, L. R. Merriweather, & J. D. Hawley (Eds.) (Vol. 2, Chap. 7, pp. 156–159).
Knowles, M. S. (1972). Innovations in teaching styles and approaches based upon adult learning. *Journal of Education for Social Work, 8*(2), 32–39.
Knowles, M. S. (1980). *The modern practice of adult education: From pedagogy to andragogy* (Revised and Updated). Cambridge, the Adult Education Co.
Koytek, P., Perin, C., Vermeulen, J., André, E., & Carpendale, S. (2018). Mybrush: Brushing and linking with personal agency. *IEEE Transactions on Visualization and Computer Graphics, 24*, 605–615.
Kumari, S., Gautam, H., Nityadarshini, N., Das, B. K., & Chaudhry, R. (2021). Online classes versus traditional classes? Comparison during COVID-19. *Journal of Education and Health Promotion, 10*, 457.
Lee, S. J., & Nuatomue, J. N. (2021). Students' perceived difficulty and satisfaction in face-to-face vs. Online sections of a technology-intensive course. *International Journal of Distance Education Technologies, 19*(3), 1–13.
Lockee, B. B., Bond, M. A., McGowin, B. M., & Blevins, S. J. (2022). Beyond design: The systemic nature of distance delivery mode selection. *Distance Education, 43*(2), 204–220.
Masdéu, M., & Fuses, J. (2017). Reconceptualizing the design studio in architectural education: Distance learning and blended learning as transformation factors. *Archnet-IJAR, 11*(2), 6–23.
Neroni, J., Meijs, C., Gijselaers, H. J., Kirschner, P. A., & de Groot, R. H. (2019). Learning strategies and academic performance in distance education. *Learning and Individual Differences, 73*(April), 1–7.
Piergiovanni, P. R. (2018). Blended learning online and hands-on activities in an applied fluid flow course. In *Blended learning in engineering education*, A. Rahman & V. Ilic (Eds.) (Chapter 2). CRC Press.
Rosson, M. B. (2014). Learning by design, In *Innovative practices in teaching information sciences and technology: Experience reports and reflections*, Carroll, J. M. (Ed.) (pp. 75–83). Springer.
Selwyn, N. (2011). 'Finding an appropriate fit for me': Examining the (in)flexibilities of international distance learning. *International Journal of Lifelong Education, 30*(3), 367–383.
Trauth, E. M., & Booth, K. (2014). Reflections on blended learning. In *Innovative practices in teaching information sciences and technology: Experience reports and reflections* (pp. 207–219).
Watermeyer, R., Crick, T., Knight, C., & Goodall, J. (2021). COVID-19 and digital disruption in UK universities: Afflictions and affordances of emergency online migration. *Higher Education, 81*(3), 623–641.

Semester Projects on Human-Computer Interaction as Service and Outreach: Undergraduate and Graduate

Frank E. Ritter and Sarah J. Stager

Abstract We present an approach for using class projects in undergraduate and graduate human-computer interaction (HCI) classes to have social and economic impact. Class projects can help teaching by being more motivating to students, help students and instructors in their careers, be more interesting to the instructor, and have more impact to society and academia. We provide two example project descriptions used at Penn State's College of IST since 2000, suggestions for how to teach project-based HCI courses, and example projects that have suggested useful interface changes to the websites of a variety of government, non-profit, university, and small-to-medium-sized businesses, as well as publishable and published graduate projects. These reports fulfill a pedagogical goal by having students demonstrate mastery of the material, particularly evidence- and theory-based changes to improve the usability of websites. The reports provide (in the best cases) strong, correct suggestions for improving the usability of these websites or extend the method, practice, and application of HCI methods. The use of these reports also helps reduce plagiarism because the sites and thus work are unique.

Keywords Project-based learning · Team-based learning · Outreach · Task analysis · Usability reports · Teaching human-computer interaction · Teaching information science and technology

Introduction

In our experience, students in the last 2 years of their undergraduate degree programs in most American, British, and German universities, participating both in residential instruction or fully remote, have enough knowledge and skills to contribute to research and engineering. They cannot do this on their own typically (although

F. E. Ritter (✉) · S. J. Stager
College of IST, Penn State, University Park, PA, USA
e-mail: frank.ritter@psu.edu; sjs230@psu.edu

J. M. Carroll (ed.), *Innovative Practices in Teaching Information Sciences and Technology*, https://doi.org/10.1007/978-3-031-61290-9_13

there are, of course, exceptions), but they can participate in guided class projects that have impact on outreach and application. Such projects not only enhance the educational experience but also develop a sense of social responsibility and community involvement among students.

In our work, we use decomposition to parse the requirements of a user-centered system. This is done by leveraging the psychology of user-centered design. Lessons on poor design found in the real world as examples of the worst interfaces allow students to develop pattern recognition of poor interfaces through applying heuristics of interface design such as those developed by Nielsen and Molich (1990), Norman (1983, 2013), Stone et al. (2005), and Shneiderman et al. (2016). Task analysis, usability testing, and other methods are also used to develop structures to improve design.

Project-based learning is supported by research showing there is improved retention and transfer of learning when exposure to content and experience is provided through project-based learning (e.g., Blumenfeld et al., 1991; Greeno & Engeström, 2005; Kokotsaki et al., 2011). Our scaffolded approach is designed for continuous and simple improvement through three key ingredients, (a) establishing a shared language, (b) empowering action, and (c) allowing time to practice with feedback. This works well for students participating in residence and fully remote.

Student work on projects is used at other places as part of clubs and, for example, at Purdue as a special projects course (EPICS), but at PSU it is applied in a standing course. Students in our human-computer interaction (HCI) courses at Penn State (IST 331, IST 413, and IST 521) use this approach. These courses are 3 credit hours that meet for 3 hours per week for 15 weeks. A normal load is 12–18 credit hours per semester. At least 2 office hours per week are provided for further consultation.

Successful student reports provide a set of suggestions about how to make systems more usable based on applying the course material. The reports have provided contributions to local companies, the university's systems, non-profits, and other organizations through a focused, real-world group project over a semester examining a website or other system interface. Similar approaches are presented by Cameron (2014) and by Bagby (2014) in the previous volume.

Where we say website, you can read this as meaning any system with an interface because the principles or concepts being used for websites are applicable to any system with a user interface. Student projects have done other HCI-related projects, examined, and, in some cases, created apps for mobile devices and skins for applications. Table 1 provides a set of example organization websites and systems that have been examined or developed in these courses.

What is novel, we believe, is the set of constraints on this approach to adding a project to courses and the scaffolding we provide. The rest of this chapter attempts to describe the approach in enough detail that you can create similar projects in similar courses.

Table 1 Example websites and systems analyzed showing the range of systems examined

Clinton Township, Wayne County, PA	[State College] Community Help Center
Security Risk Management Association	Penn State Habitat for Humanity
Listserv email list server system at PSU	Oktavamod.com (microphones)
Penn State Office for Disability Services	Your Conversation Connection
DBLP (computer science publication database)	Centre Volunteers in Medicine
PSU Undergraduate Degree Programs Bulletin	The Rowland Movie Theater
Phone app to help with depression (PsySpace)	Centre County United Way
Shakespeare's works as an app for the iPhone	PSU ITS Lab Consulting
CataBus (Centre Area Transportation Authority)	Free Software Directory
MOOC design (Wang et al., 2021)	Deihls' Flowers Inc.
Bibliography tools (Cai et al., 2021)	Northland Bowling
eLion, PSU registrar's website for students dropping classes	Campus Concierge
Southeastern Pennsylvania Transportation Authority (SEPTA)	Becky's Drive In
Walking speed diagnostic tool (Gonzalez-Vargas et al., 2021)	PSU Webmail
Keystroke logger design (Morgan et al., 2013)	Source Forge
One laptop per child usability (Yeh et al., 2010)	work.psu.edu
Banking literacy app (BancAprende.bantrab.com) for youth in Guatemala	AreUHungry
Testing architecture designs in VR (Govindarazan et al., 2023)	CiteSeer

The Class Project

The class project, which is done by a group of 2–4 students, culminates in a 10–20 page report providing 3–5 concrete, supported suggestions for changes to an interface or system. This approach does not have students simply act as HCI consultants for a fictitious interface, but to serve directly as consultants on a real interface. The suggestions may first arise from the student group's opinions of the website, but the groups are strongly told not to just complain that they do not like the website—they are required to support their suggested changes. Support for changes can come from theories, analyses, or empirical studies.

The content of these reports vary based on their project. The appropriate HCI methods to use vary based on the usability risks in each system (Pew & Mavor, 2007). Each group does the same labs, but often the methods do not apply to each system. In these cases, alternative sites or methods are used.

Material Taught and Used

The students are encouraged to use the methods taught in the course, but in some cases, groups have had to find and use additional methods or to develop modified methods. The reports are supposed to have an abstract, an introduction with a picture of the system, several analyses that lead to suggested changes, a summary (preferably with a summary table), and appropriate references to the theories and methods used.

The theories that are taught include the basic psychology of how users will interact with websites (Ritter et al., 2014): the ABCS of anthropometrics, behavioral aspects of the senses, cognition, and social factors. Suggestions to improve the interface can be based on such things as knowing how people read and interact with an interface, how they see colors, how they represent the world (mental models), how they navigate the website to find information, the limitations of memory and problem-solving, and other aspects of cognition and social behavior. They also do labs with a keystroke logger, reading on an interface, and task analysis, both creating an analysis and testing its predictions.

In some iterations of the courses using this method, before the group proposal, students are expected to individually identify an interface and complete the first stage, of proposing an interface to analyze. They are to provide a description of a proposed interface redesign including the purpose of the system (what tasks can be completed), a brief description of the expected users, and a discussion of the issues or problems with the current interface. They are encouraged to include images or screenshots to articulate each of the problems. All problems are to be mapped to or supported by the principles, heuristics, and guidelines in class readings. Each proposal is typically presented to the entire class for deliberation. This exercise allows students to share their process and form groups. The groups then extend the proposal of the selected interfaces and create a report.

Students are taught through course readings and lectures that user research is a necessity for any human-centered system design project. To gain insight into the people who will use the interface and to support student learning of how users are different from designers and developers, the next stage requires students to develop such things as user profiles, personas, and scenarios. The submission at this stage is assembled into a single, cohesive document with a brief introduction section that explains the user groups for this project.

Most reports also use a task analyses that can include time predictions. Time prediction is a part of task analysis that involves estimating how long it takes users to complete specific tasks. This metric can be crucial for assessing the efficiency of an interface. For instance, if a task takes longer than expected, it might indicate usability issues. A task analysis of an interface can usually provide suggestions for changes, including making interaction more regular across tasks or supporting important tasks more directly.

Empirical analyses involve recruiting users to test the interface. These are small studies that typically involve three to five users. Creating these studies requires discussions between the group and the teaching team about how many users to study, representativeness between the users they can recruit and the actual users of the systems, and what tasks should be done in the study. Testing the tasks on the interface, paper or digitally based, the students develop and execute usability testing. Standard usability assessments such as task time and improving the task analyses are often incorporated in this stage. Bagby (2014) notes similar problems and some types of advice for creating war rooms as student projects.

Sometimes, the point of the analysis is to explore and document how usable the interface is (including if it can be learned at all). And sometimes the analyses are used to explore or to show that a particular task is not as easy as might be expected. Discussions about the users to recruit can be very informative when done across projects. The class can see that sometimes the target users are very similar to the users that can be recruited, and sometimes the users and those that can be recruited are quite different. How to find more appropriate users to study is a useful discussion point typically shared with the whole class.

Thus, the report's suggestions are supported by published papers, analyses performed on the website, or empirical studies of the website. This should not only make the resulting suggestions more accurate but also more believable.

Prototyping Used Occasionally

The project does not require going beyond providing suggestions—revising or redesigning the website is often not possible directly, and doing so for an external site within a semester time period is difficult. Revision is difficult not only because of the direct system changes, but more often because of social and administrative processes.

For some projects, however, the group wants to make more concrete suggestions based on their review, and some projects are to create interfaces. In these cases, some groups have gotten the changes implemented on the site, but it seems inappropriate to require students to implement or get implemented changes from an organization neither the teachers nor students control. What is required, though, is being able to submit the report to someone who could make these changes, such as a webmaster, vice-president, or lab manager of the website being analyzed.

When the Adobe Suite and Creative Cloud became available as enterprise software, students gained access to the creation of digital low-fidelity and high-fidelity prototypes through Adobe XD. (Similar tools like Figma could be used.) Student groups using Adobe XD are encouraged to use the user interface (UI) kits to show what their review suggests. This tool provides an opportunity to develop no-code interactive interfaces that demonstrate imagination and iteration. Integrating tools, however, comes with the price of time to learn how to use them. In some cases, the use of paper prototyping persists. In the current volume, Gamrat (this volume) discusses and validates the use of no-tech solutions to engage learning.

If the project includes a suggested redesign, the next stage often requires each member of the design team to sketch out some initial designs on paper. These low-fidelity prototypes allow students to start the design process because they are quick and easy to create, modify, and share. No special tools or software are used, and there is no learning curve of the tool. This format allows students to individually contribute to the design and to go through several iterations quickly. They are encouraged to consult with people who are similar to the users they are designing for to get their feedback. Students then share their sketches with the team; this creates several different designs to consider. Students are then reminded that this is a method known as ideation and that the use of multiple prototypes lead to better design (Dow, 2011). The group submission requires collaboration to sketch a paper prototype that reflects a unified vision for the redesigned interface. Groups embed images of the prototype in a document and provide a brief description of each image. Because these projects vary based on the website being analyzed, the detail and number of prototypes do as well.

The next step in the project is to build a high-fidelity prototype. In the semesters that we integrate digital tools such as proto.io or Adobe XD, students use UI kits and assets to build an interactive digital form of the interface design. Submissions include a document describing the process of creating the interface and the interface including a sharable link to the design if the course integrates Adobe XD.

The Project Process

This approach and these projects follow the risk-driven spiral design model (Pew & Mavor, 2007). Thus, the groups are given feedback at several stages of the process. Groups propose a system with a title typically in Week 5 that includes a contact at the organization and then a paragraph abstract in Week 8. Sometimes they write up an outline and a one-page version around Week 12, get feedback on each analysis, and give an (ungraded) oral presentation 1 or 2 weeks before the final report is due in Week 16 (finals week).

The oral presentation allows feedback from the teaching team and joint lessons to be learned across groups. Sometimes it allows groups to collaborate or cross-cite. Many groups can use their system as the topic in a lab applying a particular method (lab homeworks are used to teach several of the methods), and groups get feedback on these lab write-ups that they can use to improve that aspect of their final report. (One early group figured this out before the first lab and made every lab a part of their final report, using feedback from each lab grade to improve the final report.) After they turn in their final report, we provide suggestions on what they should do before sharing the report with the organization and whether we encourage them. For some reports we also ask permission to add them to the website as useful examples (see Appendix), and students can point back to these reports as an accomplishment.

We are lavish with our feedback, treating the reports as early drafts that we would like to co-author, in that the reports have to have author, dates, and page numbers;

the writing has to be organized by headings; and it must look like a paper that can go outside the class because it often does. MacKenzie's (2013, Chapter 8) chapter on writing HCI reports can be useful here. Aspects of word processing that are helpful with this are also taught (Doozandeh & Ritter, 2019a, b). These aspects include how to outline a report, spell, and use paragraph styles.

To illustrate what the reports are like, we will cover a few in detail. Additional reports are available online at http://acs.ist.psu.edu/ist331/example-projects. Some are password protected because Ritter has permission to share them with the class but not further, but some have given permission to share more widely (a form granting or denying permission is now included as part of the final project submission). An example plain text form is included as an Appendix. As a group, these projects show the diversity and strength of projects.

Example 1

As an example, a group looked at the phone application, Penn State-Go, a combined graphical user interface that provides multiple online resources at Penn State/University Park. This application was chosen because one of the group members worked for the University IT service desk and as part of their job routinely responded to calls from students who needed information. Group members were interested because of the incorporation of dining menu and transportation services. Other groups have done university-based interfaces; for similar reasons, they include email list servers and work.psu.edu.

The students' report made concrete suggestions for improving the interface. The group used Adobe XD to develop assets to illustrate these changes. Some suggestions were based on task and needs analysis, including considering what information users were looking for and conducting a user study by running a preliminary survey of classmates and then modifying the high-fidelity prototype to make information easier to find by location and size.

Example 2

In IST 413, Usability Engineering, students developed an interface to respond to a need they identified from the local Society for the Prevention of Cruelty to Animals (SPCA). This group used Adobe XD, including UI kits, to create an application to responsibly dog-sit pets who are available to adopt. Their goal was twofold, to give students the opportunity to interact with pets on a short-term basis and to alleviate the pressure of caring for animals were held at SPCA waiting for adoption. The interface included images of pets available, the responsibilities required, and background checks of users. While this was a hypothetical application, the group went through steps needed to develop the resource including needs assessment, user profile, persona, scenario, prototyping and usability testing, and creating an implementable and solid high-fidelity app design.

Conclusions

This approach of applying HCI evaluation techniques to a live, real-world interfaces seems to provide motivation and a useful learning environment. Students have generally found it motivating, and the class is more fun and encouraging to teach, and there are plenty of examples from the group projects each semester to ground the topics discussed in class. They also learn how to write useful reports. These reports are useful to organizations and have also led to publications (Stark & Kokini, 2010, and projects in Table 1). The contributions of this approach have been recognized by three press releases, a university and a college, and a conference best paper.

Common takeaways reported by students participating in these course projects include:

- Improved clarity in connecting user research and testing to the design process
- The role/importance of user feedback
- The value of group collaboration in time-constrained deliverables

Basing these reports on real-world systems also helps reduce plagiarism because the sites and thus work are unique and current. We have had two groups examine the same site, but even then collaboration remained academically safe. Most sites are large enough and the study results different enough that collaboration is both worthwhile and yields distinct reports.

Most projects (about 60%) end up good enough to encourage the students to share with the organizations. About half of the other projects (20%) are good projects pedagogically, but the results are either not clear or not well presented enough to act upon. These are graded and given feedback but are not sent by us to the organizations. Some students do share them.

Discussions about all of the projects in class can lead to additional learning across groups. An exercise that we particularly recommend is asking each group to describe one good thing to do and one bad thing that they have learned from their project so far. This is usefully done about 1/2 and 7/8's of the way through the semester. If the students listen to their peers describe these lessons, they can learn not just from one project but from a room full of projects' lessons learned.

Why We Think It Works

Table 2 summarizes anecdotal beliefs about why this approach has worked. Some of these features are probably not necessary, and most are not sufficient when alone.

Table 2 Some of the features that may help this process work

1.	Group-based
2.	Lots of low-hanging fruit to improve interfaces
3.	Contact information for the system analyzed
4.	Teaching assistants helped coach the groups
5.	Reuse and revision of class exercises into the final report
6.	Motivated by their interests and sometimes support from their parents, colleagues, family, or work associates
7.	Individual sections note authors

It is probably the case that you would not need all of them to have this work in your course.

The first item is that the projects are group-based. This approach provides mini-sections for students to discuss and learn from each other and to practice their knowledge (Gergen, 1985). Groups also provide more "horsepower" to get the projects completed. Two-people groups appear to get less done than four-person groups. Also, four-person groups provide some buffer if a student is having difficulty in the class.

While the project work and groups require more support from the teacher than traditional lecture only courses, the groups can also help reduce teacher workload because they can provide some help to students in their group with a problem, including help with other class-related problems such as what was assigned, what was covered in lecture, or what will be on the test. However, groups larger than four seem to have more trouble meeting and end up with more free riding. Joint authorship but with named sections help reduce loafing.

The second item is that there is still a lot of low-hanging fruit in website and interface design. Students and faculty have no difficulty identifying systems that violate good design. The project can then document how, why, and how much they do so. There may be a relatively large disparity in this area between the ability to implement and design and basic knowledge about the material. In designing civil engineering structures, for example, the designer has to know about bridges and their environments, and students will have to know nearly as much to critique a design. In interfaces, it seems that anyone can (and does!) create a site, but many creators lack basic knowledge about users and their tasks to provide principled criticism.

The third item that helps it work is requiring contact information for the project. Some early projects looked at car manufacture's websites (e.g., Chevy and Ford). These are large, complex sites that are justifiably hard to develop, and a lot of work goes into them. There is less low-hanging fruit, and the reports would almost certainly be ignored if they are turned in without an initial contact. So, requiring a contact helps reduced the size of the site examined because the developers of the larger sites are not as available. Having a contact also means that the teacher has some help in focusing the students because the designers/contacts often have some idea of what could be improved and can help coach the students. Having a contact provides an additional and slightly unusual upside in that someone can thank them and implement the changes, which is more satisfying than just suggesting changes. Non-profits appreciate this more; universities sites do thank the students but seem to act on the supported changes less often. (These may have hidden users and uses, institutional inertia, or special security challenges, also see Ritter, 2024, and Ginsberg, 2013.) Having a contact does not mean that the contact provides a grade or has to be available at the end of the semester because grading has to be done by someone at the university and because we do not have control over the other organization.

The fourth item is the teaching assistant (TA) or assistants. There is a lot of coaching required to make these projects work. In addition to providing more resources and time periods, teaching assistants also provide a different level, perhaps a more approachable level, of support. That is, students sometimes have concerns that they would rather discuss with a TA. And, a good TA, which we have generally had, knows when to escalate concerns and when not to. That is not to say that you have to have a TA to do this type of teaching, but we attribute some success to the TAs that we have had.

The fifth item that has led to success is the reuse of class exercises in the final report. The class typically includes about four class exercises to apply an evaluation method to an interface that the group selects, not necessarily the final project interface. Allowing another interface to be used allows for groups to test other interfaces or to allow work to proceed before the class project interface is selected or to allow a method to be applied when the class project interface does not readily support that method.

These methods have included analyses of search logs, examining a learning curve in an interface, doing a task analysis, and looking at reading speed in an interface. The students do not have to use their class exercise's interface on these projects, but many groups learn that these smaller exercises are useful precursors to the final report. Not requiring use of the class project allows groups some flexibility to choose their project and to change interfaces if required due to unforeseen circumstances and if the interface does not adapt well to the class project. The groups then reuse the previous smaller projects.

Better groups also revise and sometimes extend the analyses from the labs when writing up the class project, either running a few more subjects or doing more or better analyses. In all cases they seem to improve the presentation of the work. Efforts are underway to develop a sustainable community of students who participated in these exercises. The goal of this practice is to enhance the community of people who have shared experiences and develop skills of current, future, and former students.

This reuse provides the additional teaching benefit of providing students a chance to revise their writing and work, each time for a grade. We fear that too often in current undergraduate education students do not see the writing revision process that is taught in graduate school and used in business, government, and academia. Providing formative evaluation and providing feedback are both ranked as "High" influences on learning by Hattie's (2008) meta-analysis of learning.

The sixth item, the context of the interface provides additional motivation to do well at least for one of the students in the group, but often several. If a site is chosen based on a student's avocation or family business, then at least one student is very motivated, and the others can see the impact. If the students jointly agree on a university or local site, they are all motivated in additional ways to contribute to improving that site. Having a social tie to the interface provides additional motivation beyond doing an "evaluation task" for a fictional XYZ Corp.

The seventh and final item is that individual sections of the lab and final report note authorship. This attribution helps the groups work. A common problem in group work is social loafing, and authorship of individual sections is useful when differential grading is required.

Limitations

There are limitations to this teaching approach. In contrast to a course with multiple-choice exams, this approach requires more work from the teaching team than a non-project-based course. The project aspects have to be coached at each level. This coaching is often group-specific as the method has to be applied to a novel situation, but the associated pleasure is that the teacher and student can both learn something.

There is a small risk of having students go out into the wild and run studies and interact with people outside the university. Some students lack the social skills, knowledge, interest, or time to do well on this task. Monitoring the materials and checking in with groups appears to ameliorate these risks. Also, the students can be referred to professional etiquette classes, coached directly, and referred to books on the practicalities of how to run studies (e.g., Ritter et al., 2011, 2013; Ritter & Ricupero, 2023). Also, in team projects, conflicts within teams are common and can arise from various sources, and managing these conflicts effectively could be challenging. The use of authored sections and a group contract help ameliorate but not remove these potential problems.

Finally, this approach is limited to courses that have methods that can be applied easily to real-world problems. Areas like industrial engineering may find that this approach can be applied in many courses. Courses with more theoretical material will have a harder time finding problems and applying methods.

Final Thought

This approach of using student projects is not completely innovative; the idea of doing guided professional work has been done for years by apprentices who work under supervision to perform useful activities. It is just a new kind of apprenticeship in a classroom setting with a focus on real-world application of what is taught in class.

And, at the end of the semester, rather than just turning in grades, the teaching team has the satisfaction of having helped the students do something more with their learning than just classroom exercises—the teaching team and their students end up helping their neighbors, their current and future employers, their universities, and local non-profits.

Acknowledgments John Bagby, Lisa Lenze, Erika Poole, and Changkun Zhao provided useful comments on the previous version of this chapter that helped improve the presentation. Amirreza Bagherzadehkhorasani, Hillmer Chona, Jonathan Dodge, Larissa Flade, Joy Geuenich, Zhanna Kaminska, Robert Ritter, Serhii Serdiuk, and Farnaz Tehranchi provided comments that helped this version. ACS LLC provided support for this work.

Appendix

This is the form that groups are required to return by email with their final report. It helps understand how to distribute the final report.

Name, group name, and date: ________________

Contact at the organization:

What did the contact do during the semester, how did it work?

Have you shared the report with the contact? Yes / No

Do you grant permission to Ritter to share the final report with the contact with a cover letter? Yes / No

Do you grant permission to Ritter to share the final report on the course web site? Yes / No

How likely are you to revise the report before sharing based on feedback?

References

Bagby, J. W. (2014). Cyber forensic war room: An immersion into IT aspects of public policy. In J. M. Carroll (Ed.), *Innovative practices in teaching information sciences* (pp. 117–132). Springer.

Blumenfeld, P. C., Soloway, E., Marx, R. W., Krajcik, J. S., Guzdial, M., & Palincsar, A. (1991). Motivating project-based learning: Sustaining the doing, supporting the learning. *Educational Psychologist, 26*(3–4), 369–398.

Cai, T., Chen, C., Huang, T. H. K., & Ritter, F. E. (2021). What makes a good reference manager? In *Asian CHI 21*, pp. 64–69. *Best paper award.*

Cameron, B. H. (2014). Enterprise integration: An experiential learning model. In J. M. Carroll (Ed.), *Innovative practices in teaching information sciences* (pp. 143–155). Springer.

Doozandeh, P., & Ritter, F. E. (2019a). Some tips for academic writing and using Microsoft Word. *XRDS, 26*(1), 10–11.

Doozandeh, P., & Ritter, F. E. (2019b). *Some tips for academic writing and using Microsoft Word* (Tech. Note ACS 2019-2) (6 pages). Applied Cognitive Science Lab, College of Information Sciences and Technology.

Dow, S. (2011). How prototyping practices affect design results. *interactions [sic], 18*(5), 54–59.

Gergen, K. J. (1985). The social constructionist movement in modern psychology. *American Psychologist, 40*(3), 266–275.
Ginsberg, B. (2013). *The fall of the faculty*. Oxford University Press.
Gonzalez-Vargas, J. M., Cole, C., Krishnakumar, S., Shatinsky, K., Hills, E., & Starkey, E. (2021). Innovating walking speed as a vital sign: An interface development and usability study. In *Proceedings of the human factors and ergonomics society annual meeting*, pp. 2300–2304.
Govindarazan, K., Chandran, S., & Chen, E. (2023). Impact of virtual reality navigation on emotional response to virtual architectural environments: A pilot study. In *Behavior in the built environment: Measuring, modeling, theorizing. ANFA 20th anniversary conference.*
Greeno, J. G., & Engeström, Y. (2005). Learning in activity. In R. K. Sawyer (Ed.), *The Cambridge handbook of the learning sciences* (2nd ed., pp. 79–96). Cambridge University Press.
Hattie, J. (2008). *Visible learning: A synthesis of over 800 meta-analyses relating to achievement.* Routledge.
Kokotsaki, D., Menzies, V., & Wiggins, A. (2011). Project-based learning: A review of the literature. *Improving Schools, 19*(3), 267–277.
MacKenzie, I. S. (2013). *Human-computer interaction: An empirical research perspective*. Morgan Kaufman.
Morgan, J. H., Cheng, C.-Y., Pike, C., & Ritter, F. E. (2013). A design, tests, and considerations for improving keystroke and mouse loggers. *Interacting with Computers, 25*(3), 242–258.
Nielsen, J., & Molich, R. (1990). Heuristic evaluation of user interfaces. In *Proceedings of the SIGCHI conference on human factors in computing systems*, pp. 249–256.
Norman, D. A. (1983). Design rules based on analysis of human error. *Communications of the ACM, 26*(4), 254–258.
Norman, D. A. (2013). *The design of everyday things* (Revised ed.). Basic Books.
Pew, R. W., & Mavor, A. S. (Eds.). (2007). *Human-system integration in the system development process: A new look.* National Academy Press. books.nap.edu/catalog/11893. Checked May 2024
Ritter, F. E. (2024). When you are not a stakeholder: Why unusable systems can be successes. *Ergonomics in Design, 32*(1), 56–62.
Ritter, F. E., & Ricupero, S. (2023). Running behavioral studies with human participants: Online. *Sage research methods: Doing research online. [Case study].*
Ritter, F. E., Kim, J. W., Morgan, J. H., & Carlson, R. A. (2011). Practical aspects of running experiments with human participants. In *Universal access in HCI, Part I, HCII 2011, LNCS 6765* (pp. 119–128). Springer.
Ritter, F. E., Kim, J. W., Morgan, J. H., & Carlson, R. A. (2013). *Running behavioral studies with human participants: A practical guide*. Sage.
Ritter, F. E., Baxter, G. D., & Churchill, E. F. (2014). *Foundations for designing user-centered systems: What system designers need to know about people*. Springer.
Shneiderman, B., Plaisant, C., Cohen, M., Jacobs, S., Elmqvist, N., & Diakopoulos, N. (2016). *Designing the user interface: Strategies for effective human-computer interaction* (6th ed.). Pearson.
Stark, R. F., & Kokini, C. (2010). Reducing risk in system design through human-systems integration. *Ergonomics in Design, 18*(2), 18–22.
Stone, D., Jarrett, C., Woodroffe, M., & Minocha, S. (2005). *User interface design and evaluation*. Morgan Kaufman.
Wang, J., Shen, H., Chen, C., & Ritter, F. E. (2021). Are learners satisfied with their MOOC experiences? Assessing and improving online learners' interactions. In *Asian CHI21*, pp. 215–220.
Yeh, K.-C., Gregory, J. P., & Ritter, F. E. (2010). One laptop per child: Polishing up the XO laptop user experience. *Ergonomics in Design, 18*(3), 8–13.

Knowledge Integration and Knowledge Building: Engaging with Different Perspectives to Create a Place for Knowledge Building in a Learning Environment

Mahir Akgun

Abstract This chapter explores the intersection of knowledge building (KB) and knowledge integration (KI) within educational settings, particularly focusing on a visual analytics course. The chapter discusses the challenges of incorporating KB into learning environments that primarily focus on learning outcomes. Using KI as a bridge, it illustrates the implementation of KB in a visual analytics course, highlighting the transformation from conventional learning to knowledge creation. The course implementation, involving learning and creation modules, demonstrates practical strategies for fostering knowledge building while achieving learning objectives. The chapter underscores the significance of shifting educational focus from mere information consumption to active participation in knowledge creation, thereby enriching both individual and group learning experiences.

Keywords Knowledge building · Knowledge integration · Visual analytics education · Collaborative learning

Introduction

Knowledge building is an educational approach that emphasizes the advancement of community knowledge (Scardamalia & Bereiter, 2006) and supports the type of discourse that plays a critical role in progressive idea improvement. In a knowledge building community, individuals and groups work progressively toward solving problems of interest to that community, and these solutions contribute to the community's knowledge not merely via knowledge accumulation but rather an improvement in the state of knowledge. From the knowledge building standpoint,

M. Akgun (✉)
College of Information Sciences and Technology, Pennsylvania State University, University Park, PA, USA
e-mail: mza149@psu.edu

J. M. Carroll (ed.), *Innovative Practices in Teaching Information Sciences and Technology*, https://doi.org/10.1007/978-3-031-61290-9_14

iterative idea improvement must be part of educational programs committed to idea improvement (Scardamalia & Bereiter, 2006).

In many constructivist learning environments, students are responsible for knowledge building, and they share responsibility for learning, working on knowledge problems, and negotiating ideas (Hmelo-Silver & Barrows, 2008). Knowledge building and other constructivist learning approaches (e.g., problem-based learning), however, represent different problem spaces (Scardamalia & Bereiter, 2014). Knowledge building pedagogy places knowledge creation and the advancement of community knowledge at the center of educational practice, with individual learning as a by-product (Scardamalia & Bereiter, 2006), whereas other constructivist learning approaches, such as inquiry learning and guided discovery, place knowledge creation in a conceptual space dominated by "learning." Classroom work employing constructivist learning approaches is mostly concerned with evaluating, questioning, and defending or refuting knowledge claims (Bereiter & Scardamalia, 2014). Going beyond learning to knowledge creation/knowledge building, however, requires adopting a design mode of thinking (Scardamalia & Bereiter, 2016). The design mode encompasses a broad range of activities involving exploring the idea landscape, crisscrossing it in multiple directions, identifying promising ideas, and continuously searching for a better way of doing things and for better explanations (Scardamalia & Bereiter, 2016).

Knowledge building (KB) pedagogy presents idea improvement as a core principle (Scardamalia & Bereiter, 2006); and the discourse in a KB classroom does not act as critical filter to accept or refute knowledge claims, but rather it encourages students to pursue critical inquiry to improve ideas and redirect work based on failures and advances (Scardamalia & Bereiter, 2016). Since the focus in knowledge building is on knowledge itself (Scardamalia & Bereiter, 1999), improving ideas and advancing knowledge in a classroom community may not lead to individual learning for all students (Bereiter & Scardamalia, 2014). Many educational institutions are held responsible for student success based on conventional measures of learning; thus, it is not easy for them to depart from a learning-first approach and move closer toward a knowledge building-first approach, especially if learning measurements are used to determine whether students advance to the next course in a sequence of courses. This does not mean that there is no place for knowledge building in educational institutions; but rather, because of the importance of bringing idea improvement into educational settings, it is necessary to create a place for knowledge creation in a space dominated by a focus on learning outcomes.

As an idea-centered, principle-based approach, knowledge integration (KI) is a very close neighbor to KB. KI puts learning at the center of classroom and emphasizes the importance of guiding students to engage with inquiry (Linn & Eylon, 2011). KI elicits ideas that students develop on their own and encourages students to broaden their perspective by guiding them to plan and conduct investigations to gather evidence and distinguish the best ideas among alternatives (Linn & Eylon, 2011). Because of KI's focus on learning and characteristics that are shared with KB, in this chapter, I use KI as a departure point toward KB for learning settings that place learning outcomes at the center of classroom work. My goal here is to provide insight into creating a place for knowledge building in a visualization course.

Background

Knowledge Building

The term "knowledge building" refers to an integrated framework for knowledge building pedagogies, practices, and environments (Scardamalia, 2002). As opposed to educational practices characterized by reliance on textbooks for content and memorized facts, knowledge building (KB) pedagogy places knowledge creation at the center of educational practice. The importance of knowledge generation has been key in various educational approaches such as problem/project-based learning and case-based learning. These constructivist approaches emphasize active knowledge creation in collaborative groups and transformation in the roles of students and teachers (Hmelo-Silver, 2004) in a conceptual space dominated by "learning." KB, however, emphasizes continuous work with ideas to improve them, including broad range of activities such as finding problems with current explanations, searching for better explanations, and identifying promising ideas (Scardamalia & Bereiter, 2014).

In most constructivist educational approaches, the gradual takeover of the adult's regulating role in the zone of proximal development is controlled by teachers. KB, however, promotes a shift in the takeover of executive functions (Scardamalia & Bereiter, 1991) and requires students to gradually assume higher-level cognitive responsibility for idea improvement (Zhang et al., 2009). In the case that students work with their peers in groups to solve knowledge problems, the zone of proximal development is defined by the problem-solving activities engaged in by group members, and the responsibility for controlling the zone is collectively assumed by group members. To foster collective cognitive responsibility in the classroom, students deal with top-level planning and decisions such as elaborating goals; monitoring gaps, weaknesses, and conflicts; analyzing the extent to which they have succeeded or failed; and adjusting their course of actions to meet goals (Zhang et al., 2009).

KB is a principle-based approach. Sociocultural and cognitive dynamics of knowledge creation, some of which are explained above in this section, are captured by a set of core principles. Scardamalia (2002) identified 12 principles, including real ideas and authentic problems, improvable ideas, rise above, constructive uses of authoritative sources and epistemic agency, collective responsibility for community knowledge, etc. The central goal behind these principles is to "turn high-level agency over to students and refashion classroom practices" (Chen & Hong, 2016, p. 271). As a principle-based approach, KB places the principles at the center rather than predefined tasks and procedures (Scardamalia & Bereiter, 2014) and leaves to teachers the challenge of "making adaptive classroom decisions to accommodate their different contexts and possibilities" (Zhang et al., 2011, p. 263).

Knowledge Integration as a Near Neighbor to Knowledge Building

As an idea-centric framework, knowledge integration (KI) builds upon the disconnected and fragmented ideas about science that result from students' explorations of the natural world (Matuk et al., 2016). KI encourages science learning by creating opportunities to bridge between the ideas that students bring to science class and more sophisticated ones (Linn & Eylon, 2011). Providing successful instruction plays an important role in creating such opportunities for students; and previous research has identified four metaprinciples of instruction that promote KI: make science accessible, make thinking visible, help students learn from others, and promote autonomy (Linn, 2000). The metaprinciples of KI are paired with pragmatic pedagogical principles that guide designers in creating inquiry projects in science (e.g., build on student ideas, study personally relevant problems, and scaffold inquiry). Unlike KB, the KI framework goes beyond the principles and identifies an instructional pattern (i.e., a sequence of activities performed by teachers or students) that includes four processes: elicit ideas, add new ideas, distinguish among ideas, and reflect and sort out ideas (Linn & Eylon, 2011). These processes, in combination, promote KI (Linn & Eylon, 2006) that is achieved by (1) encouraging students to articulate the ideas they have developed on their own regarding a scientific phenomenon, (2) providing students with an opportunity to add new ideas, and (3) distinguishing between these ideas (Linn & Eylon, 2011).

Collaborative inquiry is one characteristic of KI that is shared with KB; it is, however, achieved differently in KI and KB environments. KI principles along with KI processes contribute to collaborative learning by encouraging students to learn from each other (Linn & Eylon, 2011) and to make their ideas visible so others in KI classroom can build on them (Linn, 2000). Collaboration is taken into account at a different conceptual level in KB classroom where students working as a collaborative community contribute to the advancement of community knowledge and deals with high-level collective decision-making and planning (Chen & Zhang, 2016). Implementation of collaborative activities in KI classrooms emphasizes learning in collaborative groups; in KB classroom where learning takes place as a by-product, collaborative efforts of students are directed toward the advancement of community knowledge.

In the KI framework, inquiry process is guided by KI pattern and procedures; and students are encouraged to bring their science ideas to the classroom, add new ones, distinguish among the ideas in their repertoire, and plan and conduct independent investigations to sort out ideas. In KI environment, the process of distinguishing promising ideas from others and sorting out ideas appears to rely mostly on the justification mode of thinking that characterizes activities concerned with questioning, accepting, or rejecting knowledge claims. That is to say, the discourse in KI classrooms appears to act as a critical filter to justify ideas/theories. Building on each other's work is one way to work with ideas in both frameworks; while it is encouraged by KB to promote idea improvement, KI encourages students to build on other's ideas to promote learning from peers. Similarly, both frameworks emphasize the importance of idea diversity in the process, but for different reasons. In KB framework, idea diversity is conceived as essential to the advancement in the collective state of knowledge in the classroom community (Scardamalia & Bereiter, 2014), whereas KI respects diverse ideas and relies on the research showing that instructional materials designed considering diverse views held by students have more impact on student learning (Linn & Eylon, 2011).

Course Context

Purpose of the Implementation

The purpose of the implementation is to carve out a space for knowledge creation in a visual analytics course.

About the Course

The course introduces the fundamental principles, methods, and tools of visual analytics that enable security and risk analysts to synthesize information and derive insight from massive, dynamic, ambiguous, and often conflicting data. At the conclusion of the course, students are expected to (a) synthesize information and derive insight from data; (b) detect the expected and discover the unexpected; (c) provide timely, defensible, and understandable assessments using visual analytics; and (d) communicate findings and assessment effectively for action.

Challenge

Achieving these learning objectives requires the use of particular visual analytics techniques already known to scientists and professionals, including but not limited to distribution analysis, correlation analysis, time-series analysis, part-to-whole analysis, mapping data, and identifying patterns. From the perspective of knowledge creation in education (Scardamalia & Bereiter, 2022), activities designed to teach students how to use these data analysis techniques are not considered KB activities because they are designed to enable students to learn what is already known (e.g., existing visual analytics techniques), but not to engage them in the creation of new knowledge. In contrast, for activities to qualify as knowledge creation, they should be carried out to build on what is already known, advancing what is known, and producing epistemic artifacts, such as proofs, new formulations, reinterpretations, or reexplanations of others' work, generating supportive or disconfirming findings, and offering a new perspective on an issue (Scardamalia & Bereiter, 2022). Therefore, creating a place for knowledge building while achieving course objectives requires us to create an environment where students learn what is already known (i.e., enacting learning objectives for students), go beyond it, and advance the state of knowledge in the classroom community.

Implementation

To address the challenge explained in the previous section, the implementation is carried out with two modules, namely, learning and creation modules.

Learning Module

In the learning module, students are guided to complete in-class activities to learn how to use existing visual analytics techniques in synthesizing information and gaining insights from data. Table 1 presents in-class activities taking place in the learning module.

Creation Module

KB is a principle-based approach which identifies a set of principles that can guide classroom decisions in a variety of contexts. In KB environments, classroom decisions on how to translate these principles into effective classroom practices are left to teachers. In our context, KB principles informed the development of classroom practices used to bring knowledge building into the course. The following are three themes that lay out some KB principles and course practices.

Community Knowledge

KB is based on the premise that classrooms can be vibrant hubs of authentic and creative knowledge work (Scardamalia & Bereiter, 2022). This approach is not about replicating the achievements of past scholars or professionals but is focused

Table 1 In-class activities taking place in the learning module

Topic	Activities
Data types and making sense of quantities	1. Data types and central tendency measures are covered in a lecture format 2. Students are divided into different groups. Each group consists of four students. A dataset that includes variable descriptions is given to each group. Four datasets are used for this activity, and they are randomly distributed to groups 3. For each variable in the dataset, groups determine what type of variable it is and what the best central tendency is for the variable 4. Groups working on the same dataset come together to form a larger group and share their ideas on each variable in the dataset. Groups are expected to reach a consensus for each variable 5. A representative from each group presents their shared understanding of variables to the class
Analytical interaction and navigation	1. Several ways of interacting with data, such as comparing, sorting, filtering, and aggregating, are demonstrated in the class 2. Students work on an activity that enables them to interact with data in different ways, such as: Comparing values Comparing measures that were recorded at different points in time Correlation between quantitative variables Looking for interesting attributes and examining them in various ways
Analytical techniques and practices	1. Techniques and practices that can improve the effectiveness of visual analysis are demonstrated in the class 2. Students are provided a dataset and asked to gain insights from the data by using optimal quantitative scales, reference lines and regions, trellises and crosstabs, and multiple concurrent views
Analytical patterns I: visualizing distributions	1. Important concepts related to these topics, such as standardized values, distribution, trend, seasonality, and correlation were covered in the lecture format 2. Importance of using such concepts in exploring data was illustrated in the class 3. Groups were provided a dataset asked to design charts that could be used when revealing change, seeing relationships, and mapping data
Analytical patterns II: Revealing change	
Analytical patterns III: Seeing relationship	
Analytical patterns IV: Mapping data	

on generating new knowledge that contributes to the classroom community and, by extension, to society at large (Scardamalia & Bereiter, 2021). Knowledge is not confined to the mind of a single individual but rather exists across the collective cognitive landscape of the classroom. This distributed nature of knowledge means that the state of community knowledge in a classroom does not merely mirror the understanding of individual students. Instead, it reflects a more complex and synergistic collection of ideas, insights, and understandings shared and developed by the group. Groups share their ideas with other groups so that ideas can be continuously subjected to critical scrutiny for further improvement. Groups are invited to share their thoughts about other groups' ideas and propose new perspective for their fellow groups to consider. Groups are also encouraged to borrow ideas from other groups and use them to make their own ideas better. They are allowed to redirect their work by building on other groups' ideas if they find them more promising.

In the classroom community created for the implementation, groups collectively explored a dataset, generated ideas, gained insights from the data, shared them with their community (i.e., other groups in the classroom), built on other groups' ideas, and produced epistemic artifacts that were of interest to professionals beyond the classroom community, such as generating supportive or disconfirming findings and offering a new perspective on an issue. The dataset that reflects incidents of crime in the City of Los Angeles dating back to 2010 was used in the implementation. Groups were tasked to explore the dataset to understand crimes happening in Los Angeles (LA) and gain insights that the Los Angeles Police Department could act on to address potential issues. The work was authentic because groups were asked to gain insights from a real dataset and provide the Los Angeles Police Department with new perspectives about crimes that had happened. In addition, the work was creative because groups were motivated to advance the state of knowledge by offering new insights or alternative perspectives.

Sustained Idea Improvement

In KB environments, students engage not just in learning about what is already known but also in contributing to the creation of new knowledge, which shifts the educational focus from mere consumption of existing information to active participation in the knowledge creation process. It empowers students to see themselves as capable contributors to the collective intellectual capital of their community and beyond.

Students need a repertoire of approaches to inquiry, especially repeated aspects of inquiry (e.g., designing investigations, interpreting findings, making predictions, and forming arguments), in order to use investigative methods for knowledge creation. However, most of the students taking this course are new to the inquiry process. As Linn et al. (2004) emphasized, revisiting inquiry processes in classes helps students capture the reusable pattern of inquiry activities; then, students can become autonomous learners. Reduction to routinized procedures, such as reusable patterns of inquiry, leaves little room for design thinking, and so idea improvement (Scardamalia & Bereiter, 2016), but presenting reusable patterns of inquiry processes in the form of procedural scaffolds might still help us create space where students gradually take on the responsibility for idea improvement.

To place both KB framework's *idea improvement* and KI framework's *reusable pattern of inquiry* at the center of classroom work, the *elicit ideas* process along with a repeated pattern of *investigate ideas*, *revise/improve ideas*, and *reflect* was embedded into the course to support idea improvement.

- The *elicit ideas* process involves eliciting a wide range of ideas that students bring to classes (Linn & Eylon, 2011). It promotes diverse ideas by encouraging students to (1) articulate the ideas they hold about a phenomenon and (2) select a more promising idea and work toward improving it at the outset.
- The *investigate ideas* process involves conducting independent explorations and using authoritative sources in a constructive way to gather new perspectives that can be used to (a) identify relationships, differences, and abnormalities in the data and (b) understand the interplay of influential variables of interest to the classroom community. Students who work in groups are encouraged to collectively explore the data, gain actionable insights from the data, and share them with other groups. Groups can draw on other groups' findings to detect problems and inconsistencies of the ideas they work with.
- The *revise/improve ideas* process involves interpreting results, removing inconsistencies in ideas, and providing more inclusive explanations to knowledge problems. In the event that students notice that their original ideas do not look promising based on new information at hand, they are encouraged to either revise their own idea or distinguish among alternative ideas proposed by other groups and start building on a more promising one. Students are encouraged to redirect their work based on advances and failures.
- The *reflect* process involves engaging students in reflecting on progress and ideas.

To enable students to go through the processes of eliciting, investigating, improving/revising, and reflecting on ideas in the classroom, the following probing questions were employed as scaffolds in the implementation:

- *Elicit Ideas*
  - What do you think the data pertains to?
  - What questions can you think of to ask and answer?
  - Among possible questions, select one or two related questions to answer.
  - Looking at the data, what two or three variables do you find critical to the analysis that will be conducted to answer the questions you selected? What kind of variables (e.g., continuous or categorical) are they?
- *Investigate Ideas*
  - What analysis would be the best to conduct to answer the questions? For example, you may create visualizations to determine which areas of LA have the highest number of reported crimes. You may undertake the analysis to observe the patterns of these assaults. Depending on the question you want to answer, you can conduct one of the following:

    Time-series analysis
    Part-to-whole analysis

Correlation analysis
Mapping data

- *Improve/Revise Ideas*
  - Based on your own findings and those shared by other groups, do you plan to continue your exploration to gain better insights on the crime(s) you try to understand?
  - Is there a group whose insights look more promising? If yes:

    Is there anything you can take from that group and adapt into your own exploration? For example, if there is a group who gained interesting insights by focusing on a different variable that seems relevant to your exploration, you may want to consider that variable as well.
    Do you consider building on that group's idea?

- *Reflect*
  - What are strengths and weaknesses of your group's insights?

These scaffolds were used in classroom activities to provide procedural facilitation. In the implementation, students were enabled to articulate their ideas at the beginning and then go through the recurring pattern *investigate ideas*, *improve/revise ideas*, and *reflect* (see Fig. 1). Table 2 presents classroom activities engaged in by a group (henceforth, Group2).

As presented in Table 2, Group 2 went through the continuous and iterative process of refining, developing, and evolving ideas over time within the classroom community.

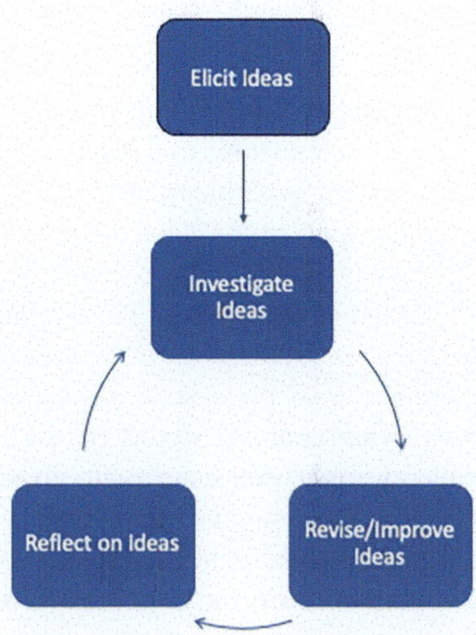

Fig. 1 Instructional process/pattern followed in the implementation

Table 2 Inquiry processes and group activities taking place in Group 2

Process	Group activities
Elicit ideas	The group expected to see a surge in firearm violence due to the increasing awareness of gun violence. They planned to identify the top ten crimes, expecting to see several crime types related to gun violence in the top ten list
Iteration I	
Investigate ideas	Group 2 conducted an analysis to identify the top ten crime types (Fig. 2). Group members decided to run a time-series analysis to determine whether there was a meaningful difference between the overall trends of crimes in the top ten list (Fig. 3). The group identified short-term fluctuations over time for most crime types. The only exception was the downward trend after 2015 for identity theft
Improve/revise ideas	During the first idea check-in session, all groups shared their insights with the class and had the opportunity to view other groups' insights. When reviewing other groups' insights/ideas, group 2 identified the variable *age* as an important variable to better understand crimes in LA, so group members decided to pursue their exploration by including the variable *age* in their analysis
Reflect	Group 2 concluded that the short-term fluctuations and the downward trend they observed may vary depending on the victims' ages. Thus, the group determined that their idea was not mature enough to end the exploration
Iteration II	
Investigate ideas	Group 2 used the variable *age* in their analysis in many ways. When the group looked at whether minors and adults had been impacted by the crime types in the top ten list differently, a notable difference between these two age groups was identified. The group determined that minors were primarily affected by incidents related to stolen vehicles (Fig. 4). The group followed up with a time-series analysis to see if the difference between adults and minors had existed for a long time (Fig. 5)
Improve/revise ideas	During the second idea check-in, all group shared their new insights with the class. Group 2 was inspired by another group that showed that communities in LA had been impacted by crimes differently. Group members focused on the possibility of gaining differing insights about how minors had been affected by vehicle theft when different areas in LA were considered. They decided to include other areas of LA in their exploration to develop a more comprehensive explanation of vehicle thefts in LA
Reflect	Ideas presented by other groups during the second idea check-in made the group less convinced about the completeness of the story they told with the data. The group figured that there was room for improvement
Iteration III	
Investigate ideas	Group 2 created several visualizations to identify similarities or differences across different areas of LA. The line chart in Fig. 6 helped them identify a spike in the number of vehicles stolen from 2014, which started declining in 2016 in almost all areas of LA
Improve/revise ideas	Group 2 figured that more context was needed to better interpret the spike between 2014 and 2016. Group members planned to enhance their understanding of vehicle thefts in LA by determining whether the observed pattern was related to other factors, such as change in population
Reflect	The group thought it would be early to talk about the comprehensiveness of their explanation

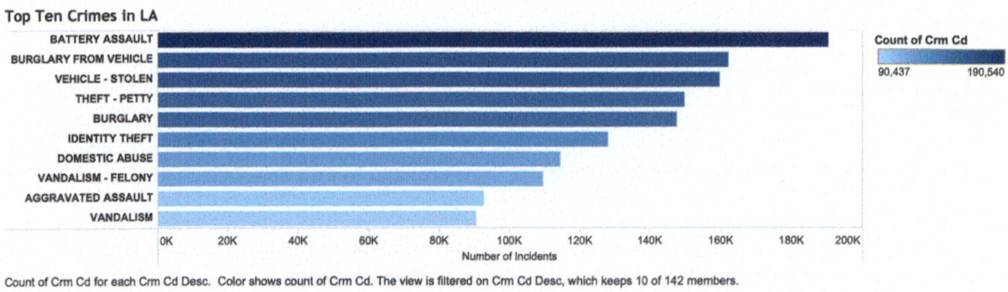

Fig. 2 Visualization I created in Iteration I, investigate ideas process

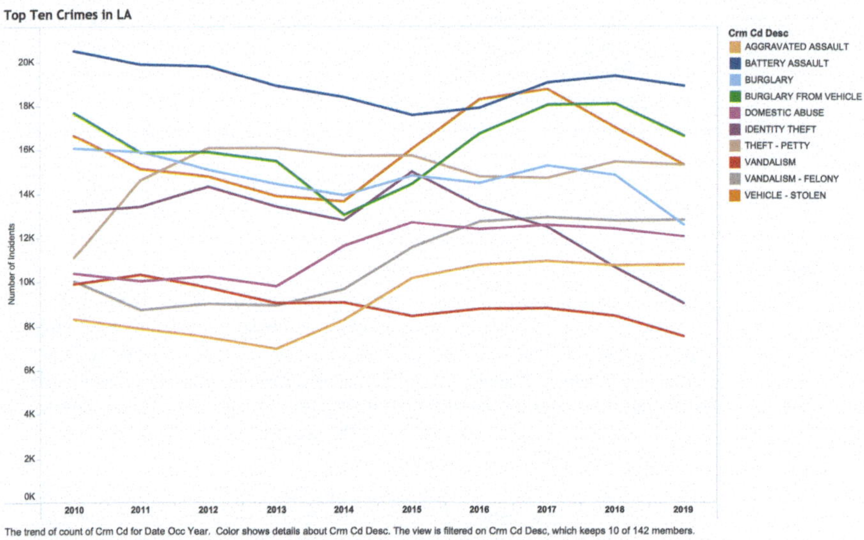

Fig. 3 Visualization II created in Iteration I, investigate ideas process

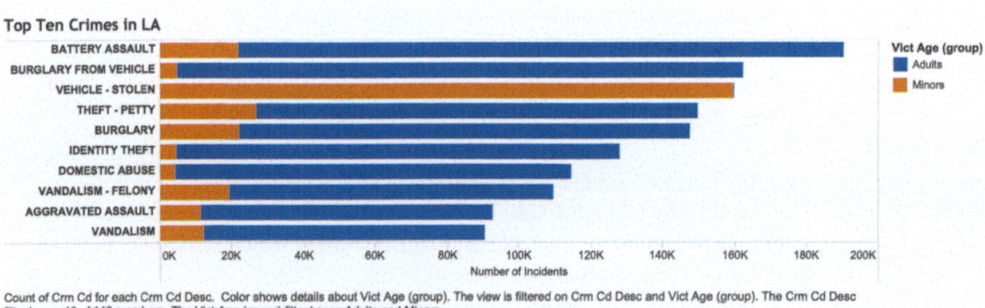

Fig. 4 Visualization I created in Iteration II, investigate ideas process

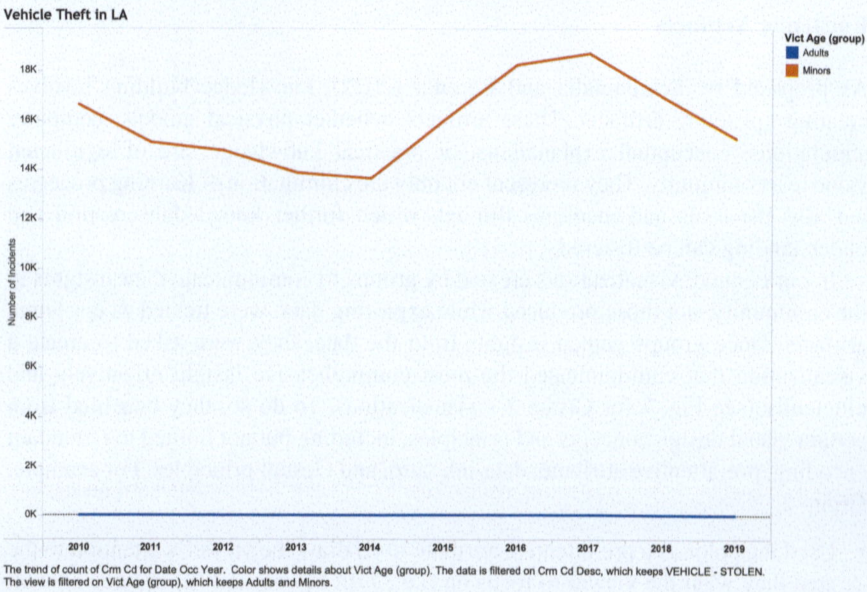

Fig. 5 Visualization II created in Iteration II, investigate ideas process

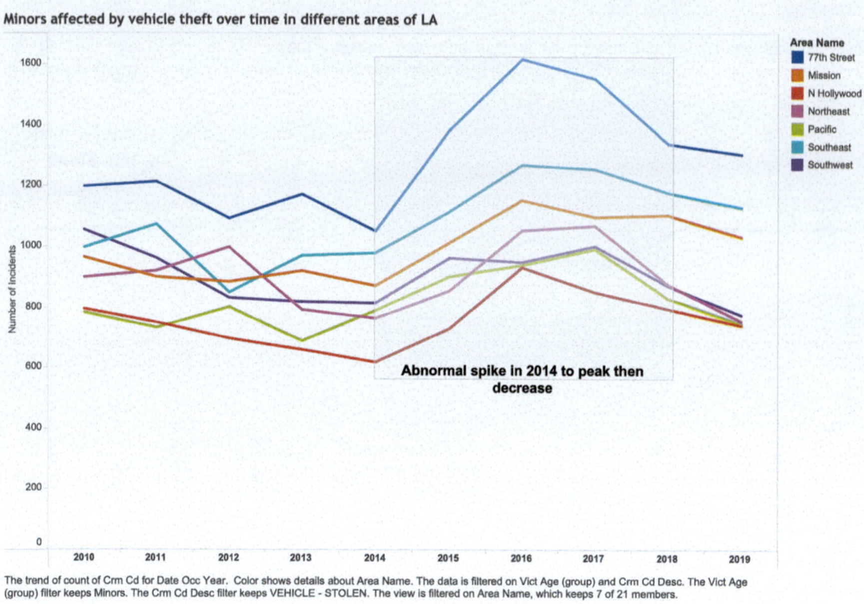

Fig. 6 Visualization created in Iteration III, investigate ideas process

Epistemic Artifacts

As proposed by Scardamalia and Bereiter (2022), knowledge building involves creating epistemic artifacts. These artifacts, whether physical models, computer simulations, conceptual explanations, or practical knowledge, are of significant value to a community. They represent not only the culmination of learning processes but also the tools and mediums through which further knowledge creation and understanding can be fostered.

In our context, visualizations created by groups to communicate their insights to the community, not those produced while exploring data, were treated as epistemic artifacts. Once groups gained insights from the data, they were asked to create a visualization that communicated the most comprehensive insight effectively and efficiently (see Fig. 7 for Group 2's visualization). To do so, they benefited from certain visual design concepts and principles, including but not limited to redundant encoding, pre-attentive attribute, data-ink ratio, and Gestalt principles. For example, Group 2:

- Used the color as a pre-attentive attribute to (i) draw the viewer's attention to the area they want the viewer to focus on (i.e., 77th Street) and (ii) provide context by graying out all other areas.
- Removed grid lines to increase the data-ink ratio.

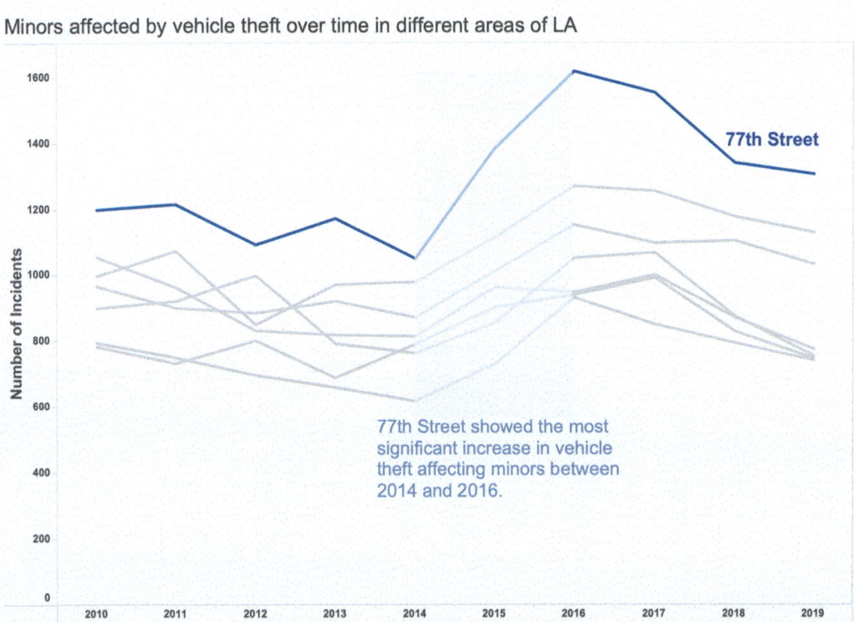

Fig. 7 Visualization created to convey the most comprehensive insight

- Benefitted from the Gestalt principle of similarity to allow the viewer to relate the trend line showing the trend for 77th Street to the text using the same color, blue.

Visualizations that communicate the message in the simplest way possible not only embody knowledge but also enable further knowledge creation and understanding. Visualizations that are easy to read and understand can help the classroom community and beyond receive the most important idea and create opportunities for them to improve it. Therefore, providing opportunities for students to learn how to use visual design principles and concepts to improve the readability of visualizations is a critical element of this process.

Concluding Remarks

This chapter has shown that it is feasible to carve out space for knowledge creation within traditional learning environments. The successful implementation of KB in the visual analytics course underscores the potential of this approach to foster engagement and idea improvement among students. By balancing the requirements of conventional learning outcomes and the aspirations of knowledge building, we can create more dynamic, collaborative, and intellectually stimulating learning environments. This approach not only benefits individual learners but also contributes to the advancement of community knowledge. Future endeavors in education should continue to explore and expand these methods, ensuring that learning environments evolve to meet the demands of a rapidly changing world where knowledge creation and innovation are paramount.

References

Bereiter, C., & Scardamalia, M. (2014). Knowledge building and knowledge creation: One concept, two hills to climb. In S. C. Tan, H. J. So, & J. Yeo (Eds.), *Knowledge creation in education* (pp. 35–52). Springer.

Chen, B., & Hong, H. Y. (2016). Schools as knowledge-building organizations: Thirty years of design research. *Educational Psychologist, 51*(2), 266–288.

Chen, B., & Zhang, J. (2016). Analytics for knowledge creation: Towards epistemic agency and design-mode thinking. *Journal of Learning Analytics, 3*(2), 139–163.

Hmelo-Silver, C. E. (2004). Problem-based learning: What and how do students learn? *Educational Psychology Review, 16*(3), 235–266.

Hmelo-Silver, C. E., & Barrows, H. S. (2008). Facilitating collaborative knowledge building. *Cognition and Instruction, 26*(1), 48–94.

Linn, M. C. (2000). Designing the knowledge integration environment. *International Journal of Science Education, 22*(8), 781–796.

Linn, M. C., & Eylon, B.-S. (2006). Science education: Integrating views of learning and instruction. In P. A. Alexander & P. H. Winne (Eds.), *Handbook of educational psychology* (2nd ed., pp. 511–544). Lawrence Erlbaum Associates.

Linn, M. C., & Eylon, B.-S. (2011). *Science learning and instruction: Taking advantage of technology to promote knowledge integration*. Routledge.
Linn, M. C., Davis, E. A., & Eylon, B. (2004). The Scaffolded knowledge integration framework for instruction. In M. C. Linn, E. A. Davis, & P. Bell (Eds.), *Internet environments for science education* (pp. 47–72). Lawrence Erlbaum Associates. https://doi.org/10.4324/9781410610393-10
Matuk, C., McElhaney, K. W., Chen, J. K., Lim-Breitbart, J., Kirkpatrick, D., & Linn, M. C. (2016). Iteratively refining a science explanation tool through classroom implementation and stakeholder partnerships. *International Journal of Designs for Learning, 7*(2), 93–110.
Scardamalia, M. (2002). Collective cognitive responsibility for the advancement of knowledge. In B. Smith (Ed.), *Liberal education in a knowledge society* (pp. 67–98). Open Court.
Scardamalia, M., & Bereiter, C. (1991). Higher levels of agency for children in knowledge building: A challenge for the design of new knowledge media. *The Journal of the Learning Sciences, 1*(1), 37–68.
Scardamalia, M., & Bereiter, C. (1999). Knowledge-building organizations. In D. P. Keating & C. Hertzman (Eds.), *Developmental health and the wealth of nations: Social, biological, and educational dynamics* (pp. 274–289). The Guilford Press.
Scardamalia, M., & Bereiter, C. (2006). Knowledge building: Theory, pedagogy, and technology. In K. Sawyer (Ed.), *Cambridge handbook of the learning sciences* (pp. 97–118). Cambridge University Press.
Scardamalia, M., & Bereiter, C. (2014). Knowledge building and knowledge creation: Theory, pedagogy, and technology. In K. Sawyer (Ed.), *Cambridge handbook of the learning sciences* (2nd ed.). Cambridge University Press.
Scardamalia, M., & Bereiter, C. (2016). Creating, criss-crossing, and rising above idea landscapes. In R. H. Huang, Kinshuk, & J. K. Price (Eds.), *ICT in education in global context: Comparative reports of K-12 schools innovation* (pp. 3–17). Springer.
Scardamalia, M., & Bereiter, C. (2021). Knowledge building: Advancing the state of community knowledge. In U. Cress, C. Rose, A. Wise, & J. Oshima (Eds.), *International handbook of computer-supported collaborative learning* (pp. 261–269). Springer.
Scardamalia, M., & Bereiter, C. (2022). Knowledge building and knowledge creation. In R. K. Sawyer (Ed.), *Cambridge handbook of the learning sciences* (pp. 385–405). Cambridge University Press.
Zhang, J., Scardamalia, M., Reeve, R., & Messina, R. (2009). Designs for collective cognitive responsibility in knowledge-building communities. *The Journal of the Learning Sciences, 18*(1), 7–44. https://doi.org/10.1080/10508400802581676
Zhang, J., Hong, H. Y., Scardamalia, M., Teo, C. L., & Morley, E. A. (2011). Sustaining knowledge building as a principle-based innovation at an elementary school. *The Journal of the Learning Sciences, 20*(2), 262–307.

Part IV
Developing Innovative Courses and Curricula

Single-Point Rubrics: An Opportunity in Graduate Education

Edward J. Glantz and Ralph J. Osmolinski

Abstract This chapter empowers graduate instructors to unlock rubric potential for deeper student learning. While tailored for graduate-level, it benefits educators of all kinds. We share our journey from tolerating to thriving on rubrics, offering insights from teaching thousands in both undergraduate and graduate settings. Key benefits include enabling self-review for students and demystifying grading. We delve into different rubric types (holistic, analytic, single-point) and guide you on personalizing and using them creatively. Real-world examples from writing-intensive capstones to hands-on cyber courses showcase application. Though focused on graduate education, valuable lessons extend to all educators seeking to enrich teaching and learning.

Keywords Single point rubrics · Graduate education · Assessment methods · Student learning outcomes · Rubric design

Introduction

We all have rubric stories to tell, and the authors' story is not much different, with one exception. For years the authors put energy into teaching, developing courses, and conducting research into teaching excellence, only to have the value and benefit of rubrics remain elusive. The authors remained unconvinced, even when auditing courses and instructors and reviewing rubric use. For example, one course included a 5 × 5 rubric with 25 cells, each including a full-text paragraph. This raised the question of whether the student would need more time to parse this rubric than to simply attempt the directions provided in the original assignment. The one exception was never giving up seeking a reasonable compromise aligning the design and use of rubrics with benefits. At the minimum, the authors hoped for a solution to

E. J. Glantz (✉) · R. J. Osmolinski
Pennsylvania State University, University Park, PA, USA
e-mail: ejg8@psu.edu; rjo102@psu.edu

J. M. Carroll (ed.), *Innovative Practices in Teaching Information Sciences and Technology*, https://doi.org/10.1007/978-3-031-61290-9_15

provide students with a more precise understanding of the criteria used to evaluate submissions. Other benefits could follow, such as providing consistency in student grading across sections and semesters, as well as expediting grading and student feedback.

This chapter begins with a general summary of the use of rubrics to enhance teaching and learning and then continues with recommendations for their emerging use at the graduate level. Graduate education, and thus rubrics, merit differentiation due to enhanced student maturity as well as teaching at higher-level Bloom's Taxonomy learning objectives.

Why Use Grading Rubrics?

We begin by sharing our belief that instructional rubrics, especially when well designed, can be valid and reliable, support learning, increase motivation, and improve efficiency for educators to provide feedback (Popken, 2020).

The authors accept that higher-education instructors in general, and those publishing rubric research, are usually quite passionate in the reception or rejection of these values (Fluckiger, 2010). Quite possibly the negative opinions center on limited previous rubric experiences, possibly with more exposure to evaluative "rating-scale" rubrics intended to expedite grading, rather than "descriptive" rubrics that support learning as well as expedite grading. This is understandable, as we also agree the former offers limited value to offset the utilization cost.

The good news is that rubric design has improved substantially, as well as the benefits from using rubrics, for both the instructor and the student. This is especially true for those teaching using a course management system (CMS) which often includes digital rubric capabilities. This chapter references examples specifically using Canvas CMS rubrics, which we suspect can be generalized across other CMS environments.

Most of us have experienced or used the densely packed **paragraph-achievement-tier rubrics**, featuring a grid of rating-scale achievement outcomes, where each cell is populated with detailed paragraph descriptions of shortcomings for each criterion. If you did not appreciate the level of detail provided by the multitude of often repetitive paragraphs, know that for several reasons, we did not either. The first is the potential difficulty for a student to parse and make sense of this rubric detail within a reasonable timeframe; the second is severely limiting the grader from providing broader constructive, more formative assessment feedback; and the third is not insignificant to tedious effort to construct these tables. These limitations become even more pronounced from the perspective of the advanced graduate student learner.

The primary benefit of paragraph-achievement-tier rubrics seems to focus on expediting the *grader's efficiency* in evaluating a limited performance set of improvement by reducing the effort to simply clicking a predefined box. Although this might be appropriate in some instances, we find this overly restrictive for most assignments at the graduate level.

Fortunately, rubric design has continued to evolve far beyond the early "analytic" paragraph rubric. Similarly, the benefits also go beyond benefiting a grader wishing to click "canned" responses. One of the major reasons to use rubrics in higher education, and often overlooked, is to provide a tool to guide the student *prior to* submission. For reference, the **single-point rubric** (SPR) clearly communicates "excellence" to the student using only a single paragraph for each criterion. Students may quickly review the requirements prior to submission. The grader still uses the rubric to show improvement needed, often supplementing the rubric with specific comments. The student is thus more clearly directed to continue improvement across a broader set of requirements more common in graduate work in subsequent assignments.

SPR fields permit instructors to provide specific feedback describing how the student exceeded, met, or fell short of achieving proficiency in each criterion. We find the flexibility and detail associated with this feature particularly appropriate in the formative assessments found at the graduate level.

Other benefits to using rubrics include:

1. Standardizing and improving grader feedback
2. Providing a tool to guide the student prior to submission
3. Improving inter-rater reliability and validity across courses and semesters
4. Connecting course learning objectives to assignment outcomes
5. Accommodating accreditation requirements to assess learning across the curriculum

We begin this chapter with an introduction to the elements common to all rubrics by describing the traditional holistic and analytic rubrics. This background is important, as all rubrics have something to offer. We continue with a more detailed introduction into the analytic-type SPR rubric and provide examples used in our graduate courses. Our hope is that the inclusion of well-designed rubrics becomes common practice across the higher education and graduate curriculum. This is somewhat selfish, as our biggest challenge is communicating to graduate students that we are using rubrics, where to find them, how to use them prior to submission, and where to find feedback information in the rubric following assessment. In the meantime, we are content simply recommending that course authors consider adopting rubrics as a "best practice."

General Grading Rubric Elements

Most educators are aware that a grading rubric is an assessment table listing in the first column various learning assignment outcome criteria. Adjacent columns list the continuum of possible achievement levels for these outcomes, ranked in terms of excellence or proficiency, often with highest from left.

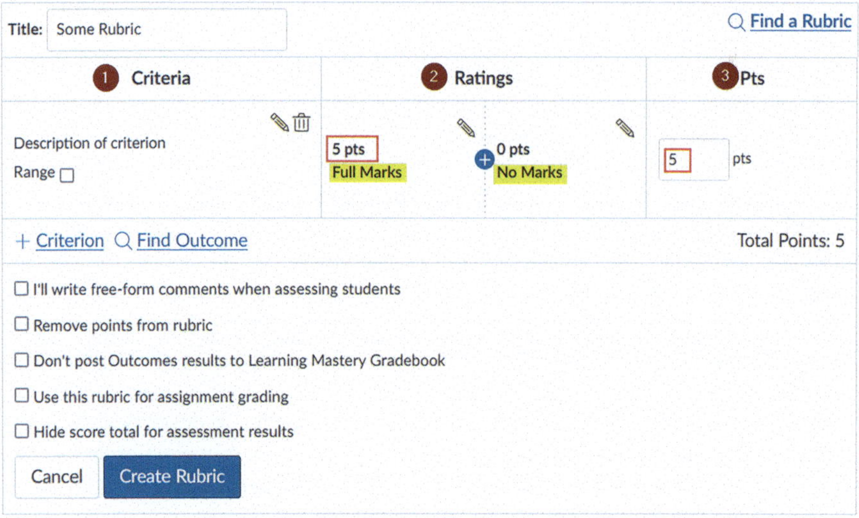

Fig. 1 Default Canvas course management system (CMS) assignment grading rubric

These elements are illustrated in Fig. 1, a default grading rubric in the Canvas course management system (CMS), illustrating:

1. *Achievement Criterion*: In the first left column, an area to describe a single learning outcome criterion is visible. Additional criteria may be added by clicking the "+ *Criterion*" link beneath the criterion cell. To edit the full text describing an outcome in Canvas, click the pencil icon in the criteria cells.
2. *Achievement Rating*: By default, Canvas rubrics are created with two achievement rating tiers. Note the default descriptive-tags (i.e., titles) used in the initial rating tiers are written "neutrally" and without judgment (e.g., full marks, no marks), as opposed to "evaluative" (e.g., excellent, poor). To increase the number of achievement tiers in Canvas, click the "+" icon in the rating cell. Clicking the pencil icon in the rating cell enables editing of the rating's score, tag, and description.
3. *Achievement Rating Score*: Each criterion's maximum rating value (e.g., 5) is displayed in the highest tier in the Canvas rating's left column and is automatically matched to the criterion's value in the right point column. When adjusting rating points in the maximum achievement left column, the value in the right point column follows. However, if the criterion's achievement value in the point column is adjusted, not only is the maximum rating increased to match, but each achievement tier is also adjusted to maintain the same normative weighted distribution (e.g., 100%, 90%, 80%, 70%, and 0%). This makes it convenient, after achievement levels have been assigned, to revise a rubric's total point value while still maintaining the weighted point distribution.

At the minimum, a rubric should provide the grader with an assessment tool to efficiently provide both *normative* and *qualitative* feedback. If provided to the student—prior to submission—the rubric becomes a tool for the student to more clearly anticipate and meet expectations. Rubrics can be modified to fit almost any type of student work, including essays, projects, presentations, and performances. Rubric outcomes can also be evaluated across the curriculum to assess student success in targeted learning outcomes (CTL, 2022).

Although rubrics may seem rather simple, there definitely is an art to designing effective rubrics. This effectiveness becomes more transparent following a granular introduction to the elements that comprise rubric design, the achievement criterion description, and the rating tiers:

Achievement Criterion Description

The educator begins with an idea of the learning outcomes from each assignment and, like course objectives, communicates this desired outcome(s) to students in writing. Like writing effective course objectives, there is a learning curve to writing effective assignment outcome criteria, which is outside the scope of this chapter. However, the benefit to educators investing time to improve objective and outcome criteria is improved course design from closing the loop between what a student *should learn* (i.e., course learning objectives) and *can demonstrate* (e.g., outcomes from the course assessments). As expected, student satisfaction will increase accordingly (Fluckiger, 2010; Reddy & Andrade, 2010).

Thus, course learning objectives and assignment achievement criterion are related. Course objectives, written using Bloom's Taxonomy language, predict the measurable knowledge, skills, or abilities that a student *should* be capable of upon course completion; the assignment's outcome achievement criteria in the rubric measures the *actual* level of student mastery in course objectives. To permit evaluation of success, assignment achievement outcomes are often more granular and detailed than related course objectives, as illustrated in this example:

1. *Course Learning Objective*—PLO 2: APPLY/CREATE: Design, develop, and apply an effective risk strategy across the enterprise.
2. *Assignment Outcome Criterion*—BLUF MEMO POST: Post contents of brief memo ("From," "To," "Subject," "Date") from risk analysis team to executive decision-maker. Section headers for Introduction (i.e., purpose of memo and roadmap of contents), Method of Analysis, Findings (i.e., top two or three hazards), and Conclusion (i.e., summary of key takeaways; no new information) sections. Cross-reference attached PDF of appendices (e.g., "In Appendix A: Divergent Analysis we show . . .").

Practically speaking, only a few distinctive outcome criteria should be used to evaluate each assignment, as too many criteria dilute the opportunity for any single criterion to have impact.

For simple "completion-only" assignments, the authors recommend one or two criteria, whereas higher-stake assignments may need as many as six. When needed, consolidate related criteria together to create a more impactful—and smaller—set of outcomes. For example, criterion describing English and grammar proficiency could be grouped into a single assignment outcome criterion called "Writing Proficiency."

Achievement Rating Levels

After drafting assignment achievement criteria, the next step is to assign *levels of achievement* to "rate" student mastery for each criterion. Educators typically do these one of two ways. The first inserts full paragraph descriptions (i.e., "paragraph-achievement-tiers") next to each achievement criterion, describing every possible achievement outcome level. The second, and recommended practice, replaces the full paragraphs with simple descriptive rating tags for each level (i.e., "descriptive-tag achievement tiers"). For example, a descriptive-tag for mastery could be, "full marks."

Although the paragraph-achievement-tier approach does give the educator a shortcut to provide student-feedback by selecting from an array of default rating options, such as "Use of Sources Criterion: 8 PTS – Only two of three credible sources cited and referenced," these paragraph-laden rubrics appear somewhat cluttered if used to guide students prior to submission. The descriptive-tag achievement-tier rubrics, instead, have the benefit of being leaner and easier for the student to understand. Either way, both paragraph and descriptive-tag rubrics include (a) full criterion description of mastery and (b) descriptive-tags with points for each rating achievement outcome tier. The authors believe "less is more" and find that by including full paragraphs for each achievement tier, the paragraph rubric diminishes the student's rubric comprehension and often oversimplifies the criteria's mastery.

For now, recognize that crafting the specific "descriptive-tag" ratings for each achievement level should follow important guidelines to not discourage students. For example, "no marks" is less judgmental than "poor." This will be discussed in more detail in the "Drafting Descriptive-Tags" section.

The authors especially recommend descriptive-tag achievement-tier rubrics if used to guide students *prior to* submission. We find *following assessment* the default statements in the paragraph rubrics too rigid and thus of limited benefit, especially at higher levels of education,

However, from a developmental perspective, we recognize that educators new to grading rubric composition may prefer to begin developing the more structured paragraph-achievement-tier rubrics as, in most cases, any rubric is better than the confusing cloud of no rubric. This is one path to develop experience to generate the true and complete description of criterion mastery required in descriptive-tag rubrics. There is nothing wrong with first becoming comfortable using rating-scale paragraph rubrics before moving on to the more formative assessment in descriptive-

Table 1 To decompose an assignment's outcomes using the paragraph-rubric rating-scale approach, use slight deviations in "repetitious" paragraphs to communicate continuity across criterion achievement levels

Achievement level / Criterion	Excellent	Acceptable	Needs improvement	No marks	Points
Introduction	**Clearly** states the main topic and provides an **excellent** preview of the essay context. 10 points	States the main topic and provides a preview of the essay context. 8 points	States the main topic **and does not** provide a preview of the essay context. 7 points	**Does not** state the main topic and does not provide a solid preview of the essay context. 0 points	___ Points

Verify a progressive flow by noting unique language in each achievement level. In this example, unique language is marked in bold text from "Excellent"

tag rubrics. Further, the effort to develop specific text describing each achievement level could help an educator unravel nuances in the assignment's outcomes.

In developing full paragraph rubrics, the authors recommend a final check for paragraph continuity across the rating tiers. To check continuity, mark unique text across achievement cells, starting with mastery. This is illustrated using bold text in Table 1, progressing from the left across achievement levels starting with "Excellent."

Table 1 also illustrates a common sequential short circuit problem using the paragraph approach. Achievements are locked in order of the cascading levels. In this example, assuming this were possible, there is no clear way to rate a student that might have provided a preview of the essay context, without adequately stating the main topic. When using the paragraph rating-scale, be sensitive to outcome diversity when locking in cascaded grading elements. This problem does not occur, however, in the descriptive-tag approach, since only the statement of mastery or proficiency is presented.

Prior to submission, the descriptive-tag rubric example in Table 2 similarly communicates the criterion mastery outcome. *Following assessment*, however, only a rating score is provided by the table. This is not a negative, as the grader is expected to, and often does at the graduate level, provide full narrative feedback. Note that both the descriptive-tag rubric (Table 2) and the paragraph-rubric rating-scale (Table 1), share the (a) criterion title, (b) full statement of mastery, (c) descriptive-tags describing each achievement level, and (d) rating score.

As the grader is expected to also provide narrative feedback, the descriptive-tag rubric is not limited to a rating score. In a modern CMS such as Canvas, supplemental feedback options include, (a) directly annotating the student's submission, (b) providing comments in the rubric criterion table, and/or (c) completing the assignment's general comments. The supplemental feedback forms in a CMS are so robust that students require direction to investigate all of them! The authors provide this information both in the assignment directions, as well as in a general comment released with the assignment score.

Table 2 The leaner descriptive-tag achievement tiers are easier for students to understand prior to submission

Achievement level / Criterion	Excellent	Acceptable	Needs improvement	No marks	Points
Introduction Clearly states the main topic and provides an excellent preview of the essay context.	10 points	8 points	7 points	0 points	____ Points

Following assessment, the rating score achievement level is supplemented with specific instructor comments

Drafting Descriptive-Tags

As "words matter," the process of drafting rubric language, including specific tags to describe achievement rating levels, is both important and challenging (Reddy & Andrade, 2010). As might be expected, descriptive-tags are frequently misunderstood and often misapplied. Most critical is to provide a concise, descriptive indicator of achievement without inferring judgment bias that could discourage students, compounding dissatisfaction upon receiving a less than perfect score. This supports the goal in higher education to provide learning on an *improvement* continuum, rather than polar positive or negative outcomes. In other words, rubrics are a powerful tool especially when used in formative feedback guiding student progress, as opposed to a summative rating assessment at the end of a marking period.

To minimize learner confusion, the authors advise using a consistent number of achievement tiers in rubrics, as well as common descriptive-tags, although the number of criteria may vary. One reason to deviate achievement tiers might be to include a completion-only criterion, such as "provide a positive peer review." Although some instructors may choose to skip descriptive-tags in the achievement tiers by only assigning point levels, the authors prefer to use descriptive-tags. The tags communicate a normative achievement level. This is especially helpful when assignments vary total points in criteria. For example, "Good" translates into 90% of the maximum criterion point value.

That words matter was not lost on educational psychologist Dr. Benjamin Bloom, who created Bloom's Taxonomy in 1956 to promote higher-level forms of thinking, as opposed to rote memorization, as well as the researchers that developed his work further to include learning "action words" (Anderson & Krathwohl, 2001). Educators are encouraged to use "Bloom verbs" to articulate measurable course and lesson objectives in cognitive processes. These processes increase in complexity from "remember," at lower levels of education, to "create," in higher. In similar fashion, the terms used to describe student achievement in mastering cognitive processes should also be thoughtful (CELT, 2022).

For example, in a simple two-tier criterion evaluation, the descriptive-tags "good" and "bad" are not suitable, as students assigned "bad" could become unnecessarily demotivated. More suitable would be to evaluate students on a progress continuum that motivates students toward improvement (Stevens & Levi, 2005, p. 41). Although research on recommended descriptive-tags for achievement tiers is limited, consider the following recommendations. Note that in each, the descriptive-tags aim to be descriptive, albeit without judgment, especially at the lower, more impactful ratings:

- Two-tier: full marks; no marks
- Three-tier: excellent; competent; needs work
- Three-tier: full marks; competent; no marks
- Four-tier: full marks; proficient; marginal; no marks
- Five-tier: full marks; good; fair; needs improvement; no marks

On a personal note, one author only uses either two- or five-level achievement tiers. The two-level achievement is useful on lower-stake assignments, often consisting of a single grading criterion (e.g., completed or not). This occurs, for example, in monitoring completion on self-introductions or reading the course syllabus and academic integrity policies. The five achievement levels align nicely with higher-stake course grading outcomes, such as 100%, 90%, 80%, 70%, and 0%.

Achievement Rating Scores

The good news is most educators naturally excel at assigning achievement rating scores, even if simply dividing total assignment point value by the number of achievement criteria. For example, a 100-point written assignment would assign 25 points to each of the educator's 4 selected criteria: organization, mechanics, reflection, and references. Of course, the final rating point distribution requires a few adjustments.

Creating a Grading Rubric

Adding rubrics to a course takes time and is subject to improvement with experience. We advise educators to get started by designing a rubric for only one high-stake assignment. Higher-stake assignments are a good place to start since they offer more recognizable achievement outcomes. Motivation to continue designing rubrics for other assignments will follow success. We also recommend deferring rubric design for lower-stake assignments until later, as these are less mission critical.

To create the grading rubric, start by assembling a rough set of grading achievement outcome criteria, along with total points for each item. Continue by fully expanding each achievement outcome criterion into a complete and measurable description, as discussed in "1. Achievement Criterion Description." This ideally describes the full accomplishment of that learning outcome, written in language understandable even to a novice reader. Although creating the initial draft takes some effort, be advised that revisions should be anticipated from outcomes observed each grading cycle. Refining assignment achievement outcomes is a work in progress.

With the developed grading outcome descriptions, the next step is dividing the total points assigned to each outcome into achievement tiers. For high-stake assignments, we recommend five-tiers, along with descriptive-tags, as previously discussed in "2. Achievement Rating Levels." Weight the tiers toward the region of grades most assigned (e.g., above 70%) and to facilitate learning use "neutral" descriptive-tags for each tier. For example, we recommended the following tags and levels: full marks (100%); good (90%); fair (80%); needs improvement (70%); and no marks (0%). This approach covers the full spectrum of possible grades, including missing submissions, and is biased toward the more frequently occurring grade assignments above 70%.

In a CMS grading rubric, although the instructor may click the rating cell to assign points for each outcome, the instructor may instead choose to insert a custom value, as well as enter a rating narrative comment for each criterion. In a Canvas CMS rubric, the narrative comment text box is enabled as shown in Fig. 1 by checking the box, "I'll write free-form comments when assessing students."

Only two-tiers might be needed in lower-stake assignments that rate binary completion criteria, such as one outcome worth 100% (full marks) and the other 0% (no marks). Again, as discussed in the "Drafting Descriptive-tag" section, these tags should be neutral (i.e., not evaluative), factual, and absent any emotion or bias that might infer judgment on the student's performance.

Note that lower-stake assignments need not be limited to one criterion. Whereas the first criterion might be, "Critical response to question prompt posted in discussion board," another criterion might simply be, "Student submits comments to another student post." For the latter "checkbox" criterion, only two outcomes are possible. Outcome tiers in higher- and lower-stake assignments are permitted to vary as needed to match the predicted outcomes. For example, if a criterion requires comments on two other introductions, three achievement tiers might be more appropriate. Achievement tiers within a rubric may vary as needed, but within reason to minimize cognitive load for the learner.

In summary, reducing variability in rubric design improves student compliance, including consistency in the number of outcomes for each achievement, the descriptive-tags, and normative point distribution. Total points per criterion, however, should vary where appropriate.

Graduate vs. Undergraduate Assessment

The information presented thus far is common to using rubrics to assess student performance, regardless of learning level. This chapter continues by focusing on rubric benefits and use in graduate education student assessment. Although rubric use in higher education is much less common than k-12, where state governments mandate assessment tracking, rubrics are only now slightly over 50% in higher education (Brookhart, 2018). Graduate educators may perceive rubrics as remedial, and possibly superfluous, especially while instructing older and more mature students, often with significant professional work experience, at higher Bloom's Taxonomy levels (e.g., analyze, evaluate, and create).

Rubrics, however, have recently begun to emerge in graduate programs. For example, at our university, the defense of doctoral and master degree research requires the student's committee to complete an analytic rubric assessing student performance across criteria. Also, accreditation organizations have begun requiring assessment of curricular performance in achieving progressive outcome improvement across course components to improve educational effectiveness. Peer teaching reviews are also based upon rubric assessment.

The authors consider rubric growth an important opportunity to provide formative classroom assessment in higher education, especially when knowledge, skills, and abilities are developed over the semester. Rubrics could provide a motivational opportunity to document formative student progress on an achievement continuum. In addition, this closes the loop by aligning degree, course, and lesson outcome objectives with actual assignment outcome assessments.

This is important to manage and sustain recent growth in master's degrees that is now outpacing awarded baccalaureate degree growth. This growth in master's degree demand stems from its status as the new "entry" credential, along with the approximate $12,000 annual wage premium. On the supply side, master's degree programs are typically less expensive for universities to offer, also with less exposure to subsidized tuition discounts. The master's degree has now successfully expanded beyond the M.B.A. degree and is also no longer seen as just a PhD "consolation prize" (Nietzel, 2018).

Rubric Types and Examples

So far, we have informally introduced the grading rubric as a grid of table information to provide student achievement feedback as well as other benefits. We described simple two-tier checkbox achievement outcomes useful for binary evaluation of low-stake completion assignments. We also described rubrics for higher stake assignments with three (or more) achievement outcome tiers. We described two general strategies to rate achievement outcomes, either the lengthier *paragraph-rubric*, or the more compressed *descriptive-tag achievement-tier rubric*.

Using this information for reference, we may now introduce the more traditional terms describing grading rubrics: **holistic** and **analytic**.

The "holistic" scoring rubric lumps all achievement criteria into a single, global assessment (Moskal, 2000). For example, the holistic mastery criterion combining clarity, organization, *and* mechanics could be, "The audience is able to easily identify the focus of the work and is engaged by its clear focus and relevant details. Information is presented logically and naturally. There are no more than two mechanical errors or misspelled words to distract the reader" (CTL, 2022, p. 1).

The rubrics discussed in this chapter, and used by the authors, are all "analytic" forms of rubrics. As opposed to holistic rubrics, analytic rubrics permit individual evaluation for each achievement criterion. The achievement criteria often appearing in the leftmost column is optionally followed by achievement tiers in adjacent columns. Cells in the achievement tiers, based on educator preference, may be empty, including a descriptive-tag, description paragraph, and/or point values. Most critical is that analytic achievement criteria are assessed independently.

The low-stake two-tier completion rubrics discussed are referred to as "check-list" rubrics, an analytic achievement criterion with binary outcome tiers, e.g., yes/no, present/absent, or pass/fail. "Developmental" rubrics are another unique analytic rubric form. However, instead of directly rating student achievement on an assignment, development rubrics rely on development theory to evaluate curricular performance (CTL, 2022).

In the next section, we will dig deeper into single-point rubrics, a more recent grading rubric innovation that we find compelling in higher education.

Single-Point Rubrics (SPR)

The single-point rubric (SPR) is yet another version of the analytic rubric form in that SPR contains multiple criteria, and each is individually assessed. In its purest form, the SPR may (1) be co-developed with students, (2) describe only proficiency, and (3) include outcome areas for instructor feedback in areas such as "Not Yet" (i.e., areas that need work), "Evidence" (i.e., how student met the standard), and "Advanced" (i.e., areas that go beyond basic proficiency). Specifically, the SPR attempts to set a goal for student achievement, but without limitations from listing pre-assigned shortcomings found in the paragraph-achievement-tier rubrics. The SPR was first introduced in a 2000 presentation but did not begin achieving recognition until published later by an attendee (Fluckiger, 2010).

We note the first SPR item, to co-develop rubrics with students, is likely at the educator's discretion. Some research, including learners-as-designers, revealed that students may not feel comfortable, nor qualified, to configure assessment outcome criteria and preferred relegating this responsibility to the instructors (Akgun et al., 2019). Quite possibly students require separate instruction in rubric design to become more comfortable. Note in the second item that the outcome is limited to describing proficiency, thus enabling the learner to creatively exceed (Popken, 2020).

The primary goal for rubrics in general, and SPR specifically, should be formative assessment, providing the learner with feedback on where they are in their learning, where they should be, and how to get there. SPR begins this learning process prior to submission by emphasizing descriptive outcome criteria for the student and following assessment in narrative remarks posted by the educator. Prior to submission, the level of performance expectation is transparent, and upon assessment the opportunity to reflect on achievement helps guide further learning (Chao et al., 2021).

The SPR use of *descriptive* criteria and outcome tags more suitably accomplishes formative assessment, as opposed to *evaluative* rating-scale language stating paragraph or source counts (Brookhart, 2018). Although evaluative rating criteria need not be completely excluded, they should be recognized as supporting grading rather than learning.

Master's Level of Rubric Assessment

The authors conducted a survey over the summer of 2022 of rubrics used across masters' courses taught online in the college. This section provides a high-level summary of the data results from the rubric evaluation. In addition, a discussion board as well as a course writing assignment rubric will be presented in the original reviewed form, along with an improvement example.

Each course received a summary score based on the *design* and *description* of sample rubrics used in the courses. *Design* refers to the appropriateness of the rubric criterion levels (rows) and performance tiers (columns). *Description* refers to the level of detail included in rubric text descriptions. Six discussion boards and six writing assignment rubrics from seven master-level courses were evaluated using a subjective scale of I (insufficient) to V (excessive), with an ideal target score of III (appropriate).

Table 3 data is from an evaluation of **discussion board** rubrics and reveals a common problem stemming from the rubric design itself. Either too few, or too many, criterion rows or tier columns are used in the rubric. In these courses, most only provided binary tiers (i.e., full or no marks columns), limiting the usefulness of the rubric for actual grading. Too few criterion rows typically resulted in long and confusing quality outcome descriptions, while too many criterions resulted in duplicate and overlapping objectives. The design problems are compounded in the text detail used in the description paragraphs, especially when an insufficient level

Table 3 Graduate course **discussion board** rubric instances rated at each level of appropriateness

	Scale Level of appropriateness				
	I Insufficient	II	III Appropriate	IV	V Excessive
Design Criterions and tiers	2	2	1	0	0
Description Text detail	1	1	1	2	1

Table 4 Graduate course **writing assignment** rubric instances rated at each level of appropriateness

	Scale Level of appropriateness				
	I Insufficient	II	III Appropriate	IV	V Excessive
Design Criterions and tiers	2	2	1	0	1
Description Text detail	1	1	1	2	1

of criterion or tiers results in too much or too little text to provide understanding for the student and grader.

Table 4 data is from the evaluation of **writing assignment** rubrics in the college masters' courses and illustrates a similarity of problems found in the discussion board rubrics. Again, most rubrics only provided binary grading tiers (i.e., full or no marks columns). Unlike discussion boards, writing assessment design frequently suffered from an excessive number of criterion rows. This problem fed into text descriptions that were excessive and thus difficult to delineate.

Discussion Board Rubric Improvement Example

Table 5 is the original form of a graduate discussion post rubric presenting several opportunities for improvement. The use of five criterion levels is normally appropriate, but in this case, the descriptions are excessive making the content difficult to understand.

Table 6 is presented as an improvement to the original form of discussion post rubric, integrating related content from different fields, breaking out the checklist of "completion of task" items as needed, and making other improvements most notably in a more suitable grading tier. The rating points are suggested for reference only. The student is presented with a clear set of instructions to achieve "excellence" in this rubric, which are derived from the discussion post's assignment. Excellence is described in the single-point rubric description column. It is recommended that the grader supplements rubric scoring with comments in the grading rubric itself as well as the general comment area. The rating column tiers align with more commonly assigned scores, although custom score assessments are permitted.

Writing Assignment Rubric Improvement Example

Table 7 is the original form of a graduate writing assignment rubric in a college Short Paper 2 assignment that presents several opportunities for improvement. The two-tier binary rating design is inappropriate since writing assignments are typically a

Table 5 Original form: discussion post rubric

Criteria	Ratings					Pts
Primary discussion post	50 pts Analyzes course concepts, theories, or materials correctly, using examples or supporting evidence	45 pts Explains and applies relevant course concepts, theories, or materials. Post does not reflect a substantive response to the discussion question	40 pts Explains some course concepts, theories, or materials. Post does not reflect a substantive response to the discussion question	35 pts Post fails to explain relevant course concepts, theories, or materials	0 pts Post is completely off topic	50 pts
Responding to classmates and instructor: responding to other's primary posts	30 pts Responds to multiple colleagues' posts. Responses are substantive and not repetitive to any other posts	25 pts Responds to one colleagues' post. Responses are adequate but not substantive. Responses are not repetitive to any other posts	20 pts Responds to one colleagues' post. Responses are not adequate or substantive. Responses are not repetitive to any other posts	15 pts Responds with short phrases, e.g., I agree, great job	0 pts Posts exact content and ideas already posted by others	20 pts
Citations	10 pts Cites at least two course readings. Applies relevant professional, personal, or other real-world experiences to extend the dialogue	9 pts Cites two course readings and contributes professional, personal, or other real-world experiences, but lacks relevance and application	8 pts Cites one course reading or contributes professional, personal, or other real-world experiences, but lacks relevance and application	7 pts Cites no readings. Does not contribute professional, personal, or other real-world experiences. Includes relevant input, but lacks application	0 pts Cites no readings. Does not contribute professional, personal, or other real-world experiences. Lacks both relevance and application	10 pts
APA: reference style and formatting	10 pts Proper APA style referencing and formatting	9 pts No more than one APA reference and/or formatting error	8 pts No more than two APA reference and/or formatting errors	7 pts More than two APA reference and/or formatting errors	0 pts Multiple issues with APA style formatting and no references	10 pts
			Total points: 100			

Table 6 Suggested revision to discussion post rubric in Table 5

Criteria	Ratings					Pts
Primary post: analyzes and incorporates course concepts, theories, or materials correctly, using examples and/or supporting evidence	50 pts Full marks	45 pts Good	40 pts Fair	35 pts Needs improvement	0 pts No marks	50 pts
Response post: responds insightfully and substantially to multiple colleagues' posts	20 pts Full marks	18 pts Good	16 pts Fair	14 pts Needs improvement	0 pts No marks	20 pts
Citations: cites (APA) at least two course readings or external sources. Applies professional, personal, or other world experiences to extend the dialogue	10 pts Full marks	9 pts Good	8 pts Fair	7 pts Needs improvement	0 pts No marks	10 pts
Reference list: provides an alphabetically sorted APA reference list at the end of the primary post	10 pts Full marks	9 pts Good	8 pts Fair	7 pts Needs improvement	0 pts No marks	10 pts
Professional writing: composes with no grammar or spelling errors	10 pts Full marks	9 pts Good	8 pts Fair	7 pts Needs improvement	0 pts No marks	10 pts
			Total points: 100			

higher-stake assignment needing more than "all or nothing" scoring. The use of five criterion levels could be reasonable, but in this case the descriptions appear jumbled making the desired outcome difficult to understand. Integrating and organizing the criterion content could reduce overlap.

Table 8 is presented as an improvement to the original Short Paper 2 rubric in Table 7, integrating related content from different fields, breaking out the checklist of "completion of task" items as needed, and making other improvements, most notably in more suitable grading tiers. The point ratings are included for reference only and align with the more commonly assigned scores. The student is presented with a clear set of instructions to achieve "excellence" in this rubric, which are derived from the paper's assignment. Excellence is described in the single-point rubric description of excellence column. It is recommended that the grader supplements rubric scoring by annotating comments directly in the student's submission, permitted in file submissions, as well as comments in the grading rubric itself and general comment area.

This review of master's degree curriculum in our college confirmed that higher education grading rubrics are somewhat uncommon. Our college has attempted to close this gap by recommending SPR rubrics as a "best practice" for course development and revision. The hope is not just to include grading rubrics but to include them with a bit more consistency and appropriateness within courses and across the curriculum.

Table 7 Original form: Short Paper 2 rubric

Criteria	Ratings		Pts
Organization: there is a definite introduction and conclusion; ideas are well organized and clearly communicated and connect tightly to central purpose	16 pts Full marks	0 pts No marks	16 pts
Strength: audience is left with a full understanding of the presenter's position; personal experience integrated where relevant and appropriate	16 pts Full marks	0 pts No marks	16 pts
Topical/subject: student demonstrates full knowledge by answering with explanations and elaboration; student supports ideas with data, evidence, examples, and facts from the course content; student includes an original idea—not a restatement of ideas posted by classmates; student has demonstrated an understanding of the topic by using the course content when drawing conclusions	30 pts Full marks	0 pts No marks	30 pts
Completion of task: identifies relevant arguments (reasons and claims) pro and con; offers analyses and evaluations of obvious alternative points of view; includes at least one correctly cited resource; engages other students and instructors for follow-up (responding to at least two other postings)	30 pts Full marks	0 pts No marks	30 pts
Mechanics/grammar: response has no misspellings or grammatical errors. Paragraph(s) is well organized and uses complete sentences	8 pts Full marks	0 pts No marks	8 pts
Total points: 100			

In addition, our best practice recommendation suggests including in assignment directions:

1. Estimated time to complete.
2. Where to find assignment's rubric.
3. Where to find instructor-provided feedback after grading.

In the Canvas course management system, supplemental grader feedback on file submission assignments can be extensive, including annotating in the student submission, comments in each rubric criterion, as well as the general comment box.

We find that grading rubrics are not yet common practice that students need trained to use SPR information, both prior to submission to review outcomes and after to review grader feedback. Prior to submission, information on Canvas rubric location is provided in the assignment. Following grade release, students are reminded to use the assignment's general comment box in Canvas of where to find the instructor's narrative feedback.

Our final comment is to expect the rubric to be a living document, subject to revision upon each grading cycle, immediately upon realizing improvement was needed to improve submission quality. The loss of context made these edits difficult to administer later. Instead, the adjustments are made immediately, although current students are not bound by compliance.

Table 8 Suggested revision to original Short Paper 2 rubric presented in Table 7

Criteria	Ratings					Pts
Professional writing/organization: no grammar or spelling errors. Professionally written (introduction and conclusion), proofed, and organized, including use of word styles for section headers. Pages are numbered. Include names only on a required title page. Figures and tables are centered on page margins, captioned (tables top-left, figures bottom-center), and each cross-referenced in the immediately preceding paragraph. Appendices are ordered by cross-reference and appear on new pages	50 pts Full marks	45 pts Good	40 pts Fair	35 pts Needs improvement	0 pts No marks	50 pts
Evidence: writing demonstrates course content understanding by providing an original idea, not a restatement, supported with course data, evidence, examples, and facts, as well as personal experience, where relevant and appropriate. Uses appropriate arguments (reasons and claims), both pro and con, and offers analyses and evaluations of obvious alternative points of view	40 pts Full marks	36 pts Good	32 pts Fair	28 pts Needs improvement	0 pts No marks	40 pts
APA in-text citation and end-of-paper reference list: Use at least one professional source, cited APA in-text and in the end-of-paper APA reference list. Optionally include useful—but uncited—references in an appendix (e.g., "Appendix B: Supplemental readings"). APA reference list is sorted alphabetically and appears on a new page at the end of the paper, before the appendices (if any)	10 pts Full marks	9 pts Good	8 pts Fair	7 pts Needs improvement	0 pts No marks	10 pts
			Total points: 100			

Getting Started Suggestions

1. *Formative assessment*: The primary goal of rubrics should be seen as formative assessment for the learner, as opposed to a simple rating scale to assist grading.
2. *Start simple*: This is a work in progress for both you, the educator, and the student. Start by developing and applying grading rubrics in one or two higher-stake

assessments. As your skills and comfort level increase, set a goal to eventually add rubrics to all assignments, even the simple, lower-stake, completion-only assignments.

3. *No surprises*: For continuity, the description of each rubric criterion outcome should be derived directly from language already provided in the original assignment directions, even if somewhat compressed.
4. *Awareness*: Inform students in the original assignment when rubrics are included. Rubrics, especially at the graduate level, are relatively uncommon and often unnoticed by students. This is particularly frustrating if feedback comments are inserted directly into the rubric.
5. *Outcome scope*: The number of criterion achievement levels should increase along with the assignment's stake. Whereas two-levels may be appropriate for completion-only low-stake assignments, five levels are more appropriate for higher-stake semester project criterion.
6. *Guide students*: Students need to be informed where to find all the grading feedback, both in advance in the assignment itself, as well as with released scores in an accompanying general comment field. Rubrics are a powerful tool, but if used, students need to be reminded to review the scores and comments.
7. *Single-point rubrics (SPR)*: We recommend using SPR rubrics. These only require development of statements describing "proficiency" in advance. Optional, but encouraged if part of the CMS, is to include neutral descriptive-tier ratings.
8. *Revision*: Rubrics in general, especially SPR, are subject to review and revision with each grading cycle. We recommend making revisions as needed, even though enforcement can be deferred.

Conclusion

Rubric use in higher education should be seen as a priority, especially in graduate education at master's level where major enrollment growth is underway. Current benefits in CMS far exceed the previous limited ability to expedite rater canned feedback by simply clicking a cell in a table. In advance of submission, the SPR rubric is a tool guiding students with achievement outcome criteria. Following assessment, the instructor provides formative feedback, guiding student learning to improve on an achievement continuum.

References

Akgun, M., Tietjen, P., Asino, T. I., & Glantz, E. J. (2019). Expanding learners as designers (LasD) in IT. In *Proceedings of the 20th annual SIG conference on information technology education (SIGITE '19)* (pp. 69–74). Association for Computing Machinery. https://doi.org/10.1145/3349266.3351409

Anderson, L. W., & Krathwohl, D. R. (2001). *A taxonomy for learning, teaching, and assessing: A revision of Bloom's taxonomy of educational objectives*. Addison Wesley Longman, Inc.

Brookhart, S. M. (2018). Appropriate criteria: Key to effective rubrics. *Frontiers in Education, 3*(22). https://doi.org/10.3389/feduc.2018.00022

CELT. (2022). *Revised Bloom's taxonomy*. Iowa State University Center for Excellence in Learning and Teaching (CELT). https://www.celt.iastate.edu/instructional-strategies/effective-teaching-practices/revised-blooms-taxonomy/

Chao, I. C., King, S., Gotch, C. M., & Roberts, M. R. (2021). Exploring the educational impact of using a single-point rubric through validation in interprofessional education. *Journal of Allied Health, 50*(4), 253–266.

CTL. (2022). *Rubrics*. Depaul University Center for Teaching and Learning (CTL). https://resources.depaul.edu/teaching-commons/teaching-guides/feedback-grading/rubrics/Pages/default.aspx

Fluckiger, J. (2010). Single point rubric: A tool for responsible student self-assessment. *The Delta Kappa Gamma Bulletin, 76*(4), 18–25. https://digitalcommons.unomaha.edu/tedfacpub/5

Moskal, B. M. (2000). Scoring rubrics: What, when and how? *Practical Assessment, Research, and Evaluation, 7*(3). https://doi.org/10.7275/a5vq-7q66. https://scholarworks.umass.edu/pare/vol7/iss1/3

Nietzel, M. T. (2018). *Five reasons why the master's degree continues to thrive*. Forbes. https://www.forbes.com/sites/michaeltnietzel/2018/12/20/five-reasons-why-the-masters-degree-continues-to-thrive/

Popken, D. (2020). *The validity and reliability of a single-point rubric to assess student writing performance*. Doctoral dissertation. Western Connecticut State University. https://repository.wcsu.edu/educationdis/94

Reddy, Y. M., & Andrade, H. (2010). A review of rubric use in higher education. *Assessment & Evaluation in Higher Education, 35*(4), 435–448. https://doi.org/10.1080/02602930902862859

Stevens, D. D., & Levi, A. (2005). *Introduction to rubrics: An assessment tool to save grading time, convey effective feedback, and promote student learning*. Stylus Publishing. ISBN-978-1-5792-2115-7.

Entering the Cybersecurity Workforce: Certification vs. College Degree

David Hozza

Abstract The volume and sophistication of cyberattacks have grown rapidly over the past 30 years. Early cyber defense was typically the responsibility of network and system administrators who lacked security-specific training and tools to detect and stop attacks. Organizations struggled to develop policies and procedures for protecting information which led to numerous attacks and outages. Risk and compliance practices did not address system vulnerabilities and were not part of the audit process. As attacks evolved the approach to cyber defense slowly started to improve. Eventually, a comprehensive approach was adopted and a new field evolved called cybersecurity.

The demand for qualified cybersecurity professionals is extremely high and I will review the situation in this chapter. The required skills are varied, making this a challenge for industry and government leaders. The data in this chapter will show that organizations are struggling to find qualified cybersecurity professionals. Educating a workforce for careers in cybersecurity is a key priority for the nation, and this work will examine several educational approaches for individuals seeking to fill entry-level roles.

Keywords Cybersecurity · Jobs · Certifications · Degrees

Introduction

In July of 2022, the National Cyber Director Chris Inglis convened a National Cyber Workforce and Education Summit at the White House. At that time, it was estimated that about 700,000 cybersecurity positions were open (The White House, 2022, p. 1). The talent shortage has made salaries and growth opportunities in this field attractive for individuals wishing to acquire the requisite skills to enter

D. Hozza (✉)
The Pennsylvania State University, University Park, PA, USA
e-mail: dhozza@psu.edu

J. M. Carroll (ed.), *Innovative Practices in Teaching Information Sciences and Technology*, https://doi.org/10.1007/978-3-031-61290-9_16

this field. Scarcity of cybersecurity talent is not a new phenomenon. In 2008, The National Institute of Science and Technology (NIST) implemented an initiative named NICE to address the skills gaps. NICE stands for the "National Initiative for Cybersecurity Education." It is a framework that categorizes cybersecurity skills by roles, competencies, and workforce skills (Petersen et al., 2020).

The motivation for writing this chapter is based on the number of questions myself, and other cybersecurity educators and practitioners receive about the best way to get into the cybersecurity field. Questions come from a variety of sources, such as prospective students, parents, and experienced professionals looking for a career change. The high salaries and number of openings are driving interest in this field. The median salary for an information security analysts in 2022 was $112,000 per year (Information Security Analysts, BLS.gov. 2023). The questions that I commonly receive fall into two categories: "Can I get a job with a certification?" or "Do I need a college degree?".

Support for experiential observations I make in this chapter are based on over 30 years of industry experience in different technology roles and my recent teaching experience. Six years ago, I transitioned from industry to a role as an assistant teaching professor in cybersecurity at Penn State University. As a teaching professor, my role is to prepare students for a career in cybersecurity. I also hold over ten certifications that I have obtained throughout my career and will shed light on those experiences.

As innovative technologies have evolved so has the demand for skilled professionals to implement and support them. Over my career, I have seen the evolution of key technological advances. Examples include open systems, networking, Internet, and cloud. The cybersecurity talent demand we have witnessed over the past few years is therefore not a complete surprise, and correlations will be drawn from those previous changes.

In this chapter, I placed emphasis on analyzing the current job openings to focus on those that fall into entry-level positions. Required skills for a common entry-level position and pathways to obtain those skills are analyzed as well.

Cybersecurity Career Opportunities

One of the first documented attempts to track down a hacker is detailed in a 1989 book named The Cuckoo's Egg: Tracking a Spy Through the Maze of Computer Espionage (Stoll, 1989). The book is considered, by some, to be a must read for cybersecurity professionals. In the book, Clifford Stoll, an astronomer-turned Unix admin, tells of his adventures tracking an intruder through a maze of computers and networks scattered around the world. Many of the hacker's intrusion attempts were on systems used by the Department of Defense, and Cliff Stoll's adventure in capturing this intruder led to the government taking computer security more seriously.

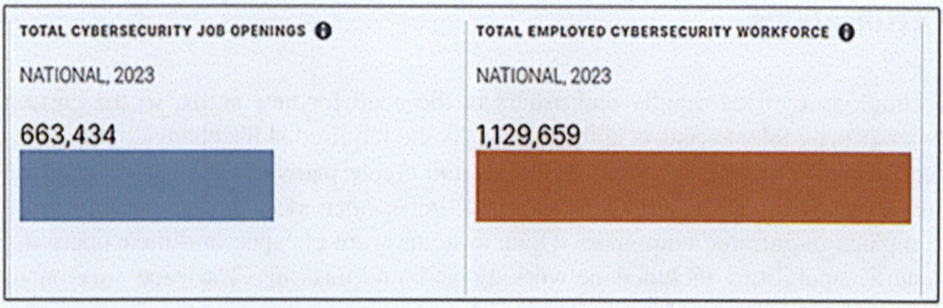

Fig. 1 Cyberseek, retrieved September 2021, https://www.cyberseek.org/heatmap.html

As mentioned in the introduction, there are approximately 700,000 cybersecurity job openings. The openings fall into roles based on the level of specialization the hiring organization requires. As the cybersecurity industry matures, these specialized roles evolve, some go away, and new roles are added. Anyone in the technology sector should always be looking several years ahead for the skills that will be required so they can stay relevant.

NIST works closely with an organization known as CyberSeek that provides employers, educators, students, and policymakers with analysis of cybersecurity job data (Cyberseek, 2017). The data illustrated below in Fig. 1 corroborates what was reported by the White House and shows the total job openings in cybersecurity and the total employed in the cybersecurity workforce.

NIST published seven high-level common cybersecurity functions (Workforce Framework for Cybersecurity, Nice Framework | NICSS, n.d.). Out of the seven NICE functions, entry-level cybersecurity candidates align with the "operate and maintain" function which is a broad category that includes seven specialty areas or job roles. CyberSeek notes that the NICE workforce categories are not mutually exclusive, so one job could perform multiple roles within the framework. The most in demand specialties from the CyberSeek heatmap data are below (Cyberseek.org Heatmap, 2023).

1. Data analyst
2. System security analyst
3. Network operations specialist
4. System administration
5. Technical support specialist
6. Knowledge manager
7. Database

Operate and Maintain Specialty Areas Of the seven specialty areas, the "cybersecurity specialist" is most aligned with entry-level positions. Job titles are not standardized like in other industries so the emphasis on cybersecurity titles is not consistent. My analysis later in this chapter will put more emphasis on combining queries with similar titles and entry-level positions:

Certifications

Technology evolves rapidly and ushers in the need for new skills, so the current cybersecurity talent crisis is nothing new. As the required skills change, it becomes necessary to retrain current technologists and create pathways for people entering this career field. For example, in the mid-1980s, open systems computing started to replace mainframe computers. Open systems were cheaper, and their operating system capabilities included networking and multitasking. The race was on to replace the mainframe. Mainframe operators and programmers either learned new skills and adapted or went the way of the dinosaur. While some banks and insurance companies still have mainframes to run legacy applications, their footprint is much smaller.

The new open system and networking technologies required skills such as Unix, Windows, and Networking administration. In the 1980s and 1990s, these skills were in high demand, and they continue to be core skills required for many of today's cybersecurity positions. Vendors like IBM, Microsoft, and Cisco started their own education departments to train existing customers on their products. Education also created a new revenue stream for these vendors and is still an important part of their offerings today. This also helped create vendor lock-in because employees certified on a vendor's specific products were invested and not likely to change vendors.

Several of these vendors partnered with skills training centers and community colleges to reach a larger audience. Boot camps, which are still popular today, were another method of training in which students are immersed in a course for a week or two, typically at the vendors' training facilities.

Certifications were either awarded for completing a training course or a competency exam and in some instances became a valuable credential. For example, in the 1990s, Microsoft's Certified Systems Engineer (MCSE) credential was a gateway to good-paying IT jobs. Today, the certification is no longer offered and was replaced with Microsoft cloud certifications. This is a good lesson that vendor-specific certifications will have a limited shelf life. For example, today's certifications from the security company Splunk will hold no value in ten years. I have a total of nine certifications from Microsoft, IBM, and EMC that are fifteen years old and no longer relevant, so I speak from experience.

Today's popular cybersecurity certifications are not vendor specific. They encompass knowledge of cybersecurity practices, operating systems, and networking. Over the past few years the certification organizations have ramped up their marketing and offerings. You should carefully evaluate any claims that a certain certification will land you a job. The analysis in this chapter will show that certifications can help, but experience and a degree are the main requirements to entering this field.

When it comes to job postings and certification requirements, the posting will often use terms like "preferred" or "good to have." Certifications are often not the barrier to entry, but they can help drive the starting salary higher. On the other side, degree requirements are usually phrased as "required." CyberSeek ranks certifications that are requested on all cybersecurity job openings, and the top four are presented in Table 1. The Certified Information Systems Security Professional (CISSP) certification is considered the top certification to have. I hold the CISSP

Table 1 Top 4 Cybersecurity certifications requested in all cybersecurity job openings, not just entry level

Certification	Openings requesting cert	Description
CISSP—Certified Information Systems Security Professional	97,555	Most request cert that requires 5 years' experience. Intense exam covering all cybersecurity domains
Security +	86,066	Foundational Cybersecurity knowledge
CISA—Certified Information Systems Auditor	75,040	Security audit skills and knowledge
GIAC—Global Information Assurance Certification	52,807	A basket of certifications in various cybersecurity domains

Cyberseek.org Heatmap (2023). Retrieved September 28, 2023, Openings Requesting Certification

certification which, in addition, to passing the exam requires 5 years of documented experience in the field. You must also be endorsed (vouched for), by another CISSP. Exam preparation can take three months to over one year to prepare for the exam. The second-place certification, the CompTIA Security plus, is a good general cybersecurity certification that is easy to obtain and intended for entry-level personnel.

Boot camps are a more intense immersive experience in cybersecurity training geared toward individuals without a technical degree. These programs are on the rise and any claims that the training leads to guaranteed employment should be vetted. Some boot camps will train you to take specific certification exams, while others train you on cybersecurity basics over a longer period such as a year. Upon completion, the student is awarded a certificate of completion. I could not find evidence in my job search data that employers were looking for certificates as a requirement to employment.

Degrees

Universities have been offering bachelor's degrees in computer science and information sciences for over fifty years. I later documented that many entry-level jobs requiring a bachelor's degree specify that it should be a technical degree like cybersecurity, computer science, or engineering.

In the past ten years, several universities and colleges have created degrees in cybersecurity. They range from associate degrees all the way to PhDs. The degree requirements for a bachelor's degree focus on core systems knowledge and cybersecurity. Penn State's program description is as follows:

The Bachelor of Science in Cybersecurity Analytics and Operations in the College of Information Sciences and Technology (IST) is an interdisciplinary program that prepares students for careers as cybersecurity professionals. It educates students on the essential concepts of cyber-defense and the analytical fundamentals of cybersecurity, with a focus on the analytical and risk management underpinnings and associated cyber-defense techniques and strategies for ensuring the safety of online information stored in large and heterogeneous networks that are embedded within and across the complex socio-technical infrastructures that are pervasive in today's business, government and military organizations. Students will acquire the knowledge and skills needed to critically assess and respond to modern information security threats, using approaches that are grounded in a holistic understanding of adversarial strategies and effective responses. More specifically, it will offer an in-depth and domain-independent approach to the development of skills in Cyberdefense technologies, tools, and processes; cybersecurity analytics and visualization; and cybersecurity risk analysis and management. The major draws from concepts and skills associated with a number of disciplines, including information science, management science, statistics and data science, human behavior, and law/policy. Graduates will be prepared to join the rapidly growing cybersecurity workforce deployed across organizations of diverse sizes and missions. (Cybersecurity Analytics and Operations, B. S. (Information Sciences and Technology) | Penn State, n.d.)

The National Security Agency (NSA) has created an accreditation mechanism known as (CAEs) or Centers of Academic Excellence in Cybersecurity (National Centers of Academic Excellence, n.d.). The University I teach at, Penn State University, has that designation. CAEs must meet the curriculum standards set forth by the program, perform community outreach, and help other organizations with solutions to cybersecurity education.

A degree program requires an investment in time and money. Students in these programs obtain knowledge of cybersecurity practices and tools augmented with interdisciplinary subjects. They develop teamwork and analytical skills sought after by many employers.

Entry-Level Career Paths

The CIA triad is one of the first information security models that a cybersecurity professional will learn. The triad is based on protecting the confidentiality, integrity, and availability of information and systems. Organizations must prioritize the assets that need protection; the things we cannot replace should be at the top. For example, the top priority is protecting people followed by information. The information is much more valuable than the servers that it resides on, servers can be replaced, but information if n protected is gone forever.

Ransomware attacks encrypt data or disable systems until an organization pays a ransom, usually in cryptocurrency, allowing the attackers to remain anonymous. These are sophisticated attacks, and few companies can recover data because they lack a robust backup system and therefore pay the ransom. According to the 2022 FBI's Internet Crimes, the FBI's crime center received 2,385 complaints resulting in adjusted losses of $34.4 million dollars (IC3 Report, 2022).

The root cause of most cybersecurity breaches in 2022 was not technical failures; it was humans. According to the 2022 Verizon Data Breaches report, 74% of security breaches involved a human element, which includes social engineering attacks, errors, or misuse (2023 Data Breach Investigations Report, 2023). Entry-level cybersecurity employees will need to understand the TTPS (tactics, techniques, and procedures) that attackers use to conduct attacks. Having staff that understand the root causes of breaches is part of the reason organizations put more value on a degree program.

The cybersecurity specialist position described earlier is an example of an entry-level position in cybersecurity. The recruiting company Indeed.com describes an entry-level cybersecurity job as one that involves the basic tasks of keeping an individual's or company's information and intelligence safe (Guide to Entry-Level Cybersecurity Job Requirements, 2023). The article outlines specific requirements as the following:

Education—Minimum is often an associate degree; however, with more education such as a bachelor's degree, they will consider less experience.
Certification—Depends on the role and the companies' value of certifications.
Experience—Most jobs require experience in a related field.
Skills—Besides technical skills, an entry-level cybersecurity job requires team players who can work in a fast-paced environment. They need problem-solving and critical thinking skills.

Another source of educational requirement data I provided in Fig. 3 from CyberSeek. Examination of openings in Fig. 2 shows that of 8,717 entry-level openings, the majority (59%) require at least a bachelor's degree. The second highest requirement is (22%) that require a graduate degree followed by (20%) for an associate degree or high school diploma (Cyberseek.org Pathway, 2023).

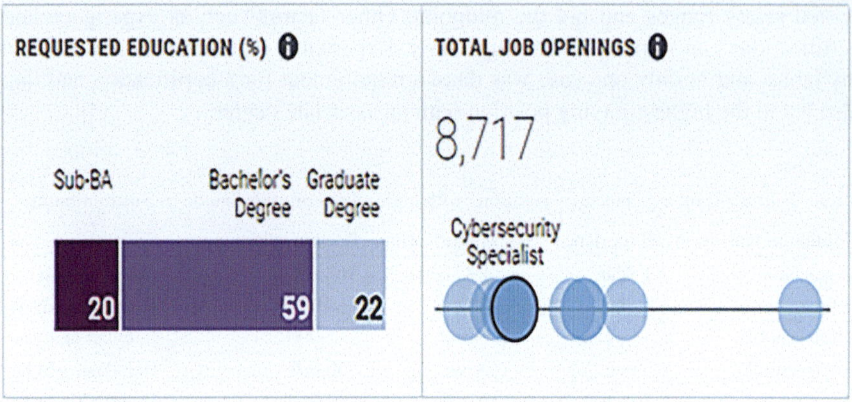

Fig. 2 Cyberseek.org Pathway, https://www.cyberseek.org/pathway.html

Fig. 3 Salaries and education, salary.com. (n.d.)

Typical Education for Entry Level Cyber Security Analyst

Degree Level	% of user with this level of education
No Diploma	7%
High School	14%
Bachelors	66%
Masters	14%

I chose to corroborate the Indeed and CyberSeek data from another source at Salary.com (Salaries and Education, salary.com. (n.d.). Figure 3 from Salary.com indicates the combination of "No Diploma" and "High School" a total of 21% which is in line with the 20% figure from CyberSeek. The bachelor's requirement in this data is higher than previous analysis.

The breakdown in Fig. 3 shows that there is a pathway to entry-level employment without a bachelor's degree, but the bachelor's is most sought by employers.

Financial Considerations

The data that I created in Tables 2 and 3 was a result of queries run on the job search site Indeed.com (Indeed, 2023). I selected cybersecurity for the keyword and then filtered on "entry level" for experience level. The education filter in Table 2 was set to high school or associate degree as a minimum. For Table 3, the education filter was set to bachelor's degree.

Since my analysis focused on entry-level positions, I used the low end of posted salary ranges and not the midpoint. Other factors such as experience and certifications can impact the starting salary. Experience requirements are listed in my tables and in only one case was there a requirement for a certification and that also led to the highest-paying position with an associate degree.

Table 2 Entry-level cybersecurity positions with high school or associate degree as minimum

Minimum education	Experience	Certification	Bottom salary range	Position
Associate	1 year	ISC certs	$62,000	Cyber analyst
Associate	None	None	$60,000	Jr. cyber analyst
High school	3+ years	None	$45,000	Jr. cyber analyst
High school	None	None	$50,000	Helpdesk—SOC
High school	1 year or degree	None	$42,000	Helpdesk—SOC
				Average $51,800

From Indeed.com, retrieved September 28, 2023

Table 3 Entry-level cybersecurity positions with bachelor's degree as minimum

Minimum education	Experience	Certification	Bottom salary range	Position
Bachelor's	1 yr preferred	None	$83,000	Cyber analyst
Bachelor's	None	None	$75,000	Info sec analyst
Bachelor's	None	None	$79,000	Security engineer
Bachelor's	1 yr preferred	None	$60,000	Info sec analyst
Bachelor's	None	None	$71,000	Cyber analyst
				Average $73,600

From Indeed.com, retrieved September 28, 2023

Observations from the job posting tables:

- The average starting salary with a bachelor's was $73,600 vs. $51,800 without.
- Bachelor's degree requirements on all posts that I examined had to be in cybersecurity-, engineering-, or computer science-related fields.
- Only one position required a certification.

The average cost of a bachelor's degree varies from private to public universities and state. Candidates need to consider the return on their investment. Based on cybersecurity job data, the return can be substantial over a career. The number of openings in this field is expected to grow so the likelihood of downsizing in this field is low which should factor into the return on investment (ROI) analysis of which entry path to take.

Summary and Conclusion

In this chapter, I have shown that the demand for cybersecurity professionals is high with almost 700,000 open positions and a median annual salary of $112,000. Education requirements for entry-level positions showed that 20% did not require a bachelor's degree but did require at least one year of experience and/or an associate degree. The average entry-level salary for people without a bachelor's was $51,800 as opposed to $73,600 with the degree.

Positions advertised for nondegree holders fit into the category of helpdesk or security operations (SOC) roles, which are traditionally lower-paying positions requiring less education. The degree holder positions, including associate degrees, consisted mostly of analyst roles which is a more senior position. Certifications did help to improve starting salaries of degree and nondegree holders. However, I did not find evidence that a certification or boot camp certificate was the only requirement for an entry-level position. Also keep in mind that certificates expire, whereas degrees do not. Degrees can open the door for advancement to senior-level positions and a higher salary.

Certifications and boot camps can be a reasonable option for someone already in the IT field looking to move into a cybersecurity role. Members of the military that acquired technical skills during their service can also leverage certificates and boot camps to enter the field. Universities also offer online degree programs for industry professionals and military members to earn cybersecurity degrees while working; the military and some employers offer tuition assistance as well. Claims made by boot camps and universities about job placement should be vetted before committing the time and money to them. Find out about their career services and where graduates are working, how much they are making, and any alumni networking opportunities.

The decision on how to secure an entry-level position is based on your own personal situation. Hopefully, this chapter provided the opportunities and options to help make your decision.

References

2023 Data Breach Investigations Report (2023). *Verizon Business*. [Online]. Available: https://www.verizon.com/business/resources/reports/dbir/

Cyberseek.org Heatmap (2023). [Online]. Available: https://www.cyberseek.org/heatmap.html

Cyberseek.org Pathway (2023). [Online]. Available: https://www.cyberseek.org/pathway.html

Cyberseek (2017). *NIST*. [Online]. Available: https://www.nist.gov/itl/applied-cybersecurity/nice/cyberseek

Guide to Entry-Level Cybersecurity Job Requirements (2023). [Online]. Available: https://www.indeed.com/career-advice/finding-a-job/entry-level-cyber-security-jobs-requirements

IC3 Report (2022). *FBI*. [Online]. Available: https://www.ic3.gov/Media/PDF/AnnualReport/

Indeed (2023). *Indeed.com*. [Online]. Available: https://www.indeed.com/

Information Security Analysts (2023). [Online]. Available: https://www.bls.gov/ooh/computer-and-information-technology/information-security-analysts.htm

Petersen, R., Santos, D., Smith, M. C., Wetzel, K. A., & Witte, G. (2020). Workforce framework for cybersecurity (Nice framework). *National Institute of Standards and Technology*. https://doi.org/10.6028/NIST.SP.800-181r1

Stoll, C. (1989). *The cuckoo's egg: Tracking a spy through the maze of computer espionage* (1st ed.), Doubleday.

The White House (2022). Announcement of white house national cyber workforce and education summit. *The White House*, July 18, 2022. [Online]. Available: https://www.whitehouse.gov/briefing-room/statements-releases/2022/07/18/announcement-of-white-house-national-cyber-workforce-and-education-summit/

What is Cybersecurity? | CISA (2021). [Online]. Available: https://www.cisa.gov/news-events/news/what-cybersecurity

Teaching as the Practice of Cultivating Relationships

Priya C. Kumar, Yao Lyu, and Ananya Reddy

Abstract Most discussions about teaching focus on how to achieve the goal of learning. But teaching is also a practice of relationship-building that thrives when meaningful bonds form among its participants. In this chapter, we (an instructor, teaching assistant, and student) illustrate how educators in IST can center relationships in their teaching practice, using examples from the course that brought us together. Priya and Yao have taught the course, an upper-level undergraduate elective on technology and society, for five consecutive semesters, and Ananya was a student in their third iteration. We walk through four principles of relationship-rich education, offering details about how the course's assignments, activities, and policies embody these principles. For each principle, we also offer personal reflections that bring to life the way these course design decisions affected us in our own educational journeys. We conclude with observations about how instructional teams and administrators can foster relationship-centered learning environments on their campuses.

Keywords Relationship-centered teaching · Student-centered teaching · Pedagogy of kindness · Self-reflection · Teacher professional development

Introduction

Most discussions about teaching focus on its function as an educational practice. And with good reason, considering that learning is the stated goal we strive toward. But as the COVID-19 pandemic brought into glaring relief, sites of schooling are also places of vital social connection. Indeed, "[r]elationships are essential because there is no learning without relationships" (Bass, quoted in Felten & Lambert, 2020, p. 1). Thus, teaching is also a relational practice that thrives when meaningful bonds

P. C. Kumar (✉) · Y. Lyu · A. Reddy
Pennsylvania State University, University Park, PA, USA
e-mail: priya.kumar@psu.edu

J. M. Carroll (ed.), *Innovative Practices in Teaching Information Sciences and Technology*, https://doi.org/10.1007/978-3-031-61290-9_17

form among instructors, assistants, and students and when its participants engage in self-reflection on their own evolving identities (Schwartz, 2019).

We three authors occupy different roles in the educational context. Priya is an assistant professor, Yao a doctoral candidate, and Ananya an undergraduate student in IST. Priya and Yao have taught an upper-level undergraduate IST course on technology and society for five consecutive semesters, with Priya as instructor and Yao as teaching assistant. Ananya was a student in the third iteration of the course. The course brought us together, but our relationships have gone beyond it: Priya is a member of Yao's dissertation committee, and Ananya is leading a study as a research assistant with Priya. We recognize that such longevity is rare and that most classroom relationships fade once the term ends. Nevertheless, attending to relationships not only enriches the teaching and learning experience (Schwartz, 2019), but it also helps students learn more and complete their degrees (Hagenauer & Volet, 2014; Mayhew et al., 2016).

In this chapter, we illustrate how educators in IST can center relationships in their teaching practice, using the example of the course that brought us together. First, we briefly explain what the course covers and how it is structured. We then walk through four principles of relationship-rich education (Felten & Lambert, 2020), offering details about how the course's assignments, activities, and policies embody these principles. For each principle, we offer personal reflections that bring to life the way these course design decisions affected us in our own educational journeys. We conclude with observations on how instructional teams and administrators can foster relationship-centered learning environments on their campuses.

Course Structure

The course, titled "The Information Environment" (IST 431), is an upper-level elective taken by a mix of third- and fourth-year students from IST, business, and communications majors. It explores how social, cultural, economic, and political forces shape digital technologies and how digital technologies in turn shape those forces. The goal of the course is to change the way students think about technology by encouraging them to approach technologies not as inert tools, but as dynamic systems.

The first half of the course focuses on the dynamics that shape the information environment (e.g., technosolutionism, design, labor) and the second half on challenges facing the information environment (e.g., surveillance, misinformation, media manipulation). Each week has a dedicated topic, with assigned readings and a reading guide that briefly explains how the readings connect to the course. The course meets twice a week, typically Tuesday/Thursday. The Tuesday sessions begin with a student group-prepared presentation introducing the week's topic, followed by lecture and group discussion. The Thursday sessions involve group activities where students apply that week's concepts to a given scenario, and each group informally presents their conclusions to the class.

Assignments are writing intensive. At the beginning of each week, students submit a 500-word memo responding to a prompt related to the readings. In-class activities involve individual and group writing. In lieu of a final exam, students write a 10-page paper on a topic of their choice. The final paper assignment is separated into components that are spread throughout the course: a roadmap, first draft, peer review, and final draft. The aim of the assignments is to help students develop the skills of identifying core takeaways from readings, analyzing situations to inform decision-making, and crafting arguments using external evidence. The in-class activities, including discussion, group work, and presentations, offer opportunities for students to gain comfort and experience communicating their ideas.

Fostering a Relationship-Rich Course Environment

Students enjoy a relationship-rich education when they (1) "experience genuine welcome and care," (2) are "inspired to learn," (3) "develop a web of significant relationships," and (4) "explore questions of meaning and purpose" (Felten & Lambert, 2020, pp. 17–18). While students bear primary responsibility for developing relationships and engaging in their education, instructors play a critical role in fostering relationship-rich education (Felten & Lambert, 2020). This section explains how IST 431 embodies each principle.

Genuine Welcome and Deep Care

Before students can experience the academic, social, and personal transformations of university education, they must feel like they belong (Felten & Lambert, 2020). Instructors can instill a sense of belonging by demonstrating that they value students as people, rather than as brains to be filled with knowledge or bodies to be disciplined through institutional policies (Schoem, 2017). The teacher-student relationship, like many relationships, is nuanced and context-dependent, but instructors can foster a strong one by laying a foundation of care, setting clear expectations, and communicating openly (Hagenauer & Volet, 2014).

The first day of IST 431 begins with a round of introductions, with Priya, Yao, and students each sharing what they go by (name, pronouns, whatever each person feels comfortable sharing), their year and major, where they consider home, and their response to an icebreaker question: "What is an app that you find interesting or enjoy using?" The question gives students an opportunity to share an aspect of their daily life that is directly connected to course content. Priya mentally notes student responses and uses the apps they mention in examples throughout the course. Students also make a name tag, which they use for the first few weeks until Priya has learned everyone's names. During the first week, students also fill out a "Getting to know you" questionnaire that asks:

- What do you hope to learn from this course?
- What kinds of careers are you interested in? If you don't know, that's OK!
- What obstacles might prevent you from succeeding this semester? Is there anything about the course specifically that concerns you?
- How can I [Priya] help you succeed this semester?
- Is there anything else you'd like me to know about you?

Priya reads and responds to each submission in the learning management system, offering advice and inviting students to meet with her or Yao if they would like to discuss any of their concerns. Opportunities for holistic feedback continue through the semester. Per the university's progress reporting policy, in Weeks 3 and 7, faculty receive automated messages with direct links to the campus advising system. Priya leaves a written comment for each student, complimenting their effort, offering advice, and/or encouraging them to meet with her or Yao to discuss any issues. In Week 8, students fill out a mid-semester survey that asks:[1]

- What is one thing you've learned in this class that really surprised you? (The more specific you can be the better.)
- What is one thing we've discussed in this class that you wish you understood better?
- Which has been your favorite week?
- With regard to teaching in this course, what would you like me to *continue* doing?
- What would you like me to *start* doing?
- What would you like me to *stop* doing?
- Thinking about your own learning, what steps can you take to be successful in this course?

Priya then summarizes their comments and presents them to the class. For instance, in the first iteration of the course, students appreciated the variety of opportunities to learn from their classmates through group work, class discussion, and formal group presentations, though they also wanted more opportunity to reflect before sharing thoughts with the whole class. Students also acknowledged that they could push themselves to contribute more to class discussions. So, Priya committed to creating more opportunities for reflection but also invited the students to participate, not to please her but because their classmates wanted to hear from them.

During the last week of the semester, Priya gives students class time to fill out the university course evaluation, which includes two numerical questions that ask students to rate the course and instructor and two open-ended questions that ask students what aspects of the course helped them learn and what aspects they would change. She invites their honest feedback, especially related to the readings, assignments, and structure of class time, but also requests that they provide specific

[1] These questions come from Badia Ahad-Legardy (Ahad-Legardy & Ocampo, 2021) and Chris Gamrat (personal communication).

examples in their open-ended responses and explain why various elements of the class did or did not work for them. She explains that she reads every response and uses it to inform future iterations of the course. Concurrently, IST has an evaluation process for TAs, where students and instructors fill out a survey providing feedback in the middle and end of the semester. While online surveys are certainly mechanistic and bureaucratic, Priya seeks to humanize such processes by providing individual responses and by explaining why such information is collected and how it is useful. Her aim is to demonstrate that student perspectives matter and can shape the teaching and learning process.

While feedback surveys focus on holistic or metacognitive dimensions of the course, Priya also tries to make it as easy as possible for students to manage day-to-day course responsibilities. All course readings and homework assignments, with deadlines, are posted in the course learning management system at the beginning of the course. The top of the course learning management system has a list titled "Resources for success in this class," which includes links to the syllabus, a research guide for the final paper, an illustrated guide for avoiding plagiarism, tips on effective reading, note taking, managing tech use in class, and scheduling meetings with the campus writing center, each of which Priya discusses during the first week of the semester and references throughout the course. At the end of each Thursday class, Priya presents a slide that reminds students about what upcoming assignments are due and when. Priya also consistently encourages students to contact her before assignments are due if they envision problems meeting the deadline, and she offers extensions when needed. Additionally, the two lowest memo scores and four lowest in-class activity scores are dropped, which gives all students flexibility to miss a few assignments for any reason.

If Priya notices a student consistently missing class or not submitting assignments, she emails them to check if everything is OK. She is careful to focus the message on their well-being rather than grades and offers to work with them if they are experiencing extenuating circumstances that affect their ability to perform in class. While students don't always respond, those that do have shared dealing with mental and physical health challenges as well as significant family caregiving responsibilities. Some take Priya up on her offer to make up missed work, while others do not, which remains their prerogative. While Priya's role as an instructor ultimately requires her to evaluate students on what they complete (or don't) during the course, she aims to ensure that every student knows they have an open invitation to work with her to make the course as successful a learning experience as possible.

Reflection from Ananya: *From my experience, building a rapport and meaningful two-way conversation with students is not straightforward. It takes conscious effort from both sides. IST 431 was one class where the effort was apparent. The teaching team was able to create an environment that emphasized and thrived on having conversations with and learning from one another, making it one of my most memorable learning experiences in college.*

In the first class, I remember finding the course structure unique in the amount of group work it included. Every week we had at least two in-class activities that involved working in pairs or in small groups. Group work was always deeply rooted in what we were learning that week and consequently felt like a welcoming opportunity to discuss our own thoughts and ideas while we were still in the classroom. I find myself pre-occupied with my own ideas on a subject during class if I am not given the chance to voice my thoughts out loud. While I was aware that group work still counted as an assignment, the deliverables felt low stress as they tended to involve short text entries or informal presentations. Above all, we were encouraged to have meaningful and collaborative conversations with one another. Being a student that mostly thrives in individual work, I found it surprising how much I enjoyed the group discussions. I would also attribute this comfort to how Priya and Yao seemed to greatly value the opinions presented by the class, creating a meaningful two-way conversation.

Reflection from Yao: *In the first semester working with Priya, I considered myself as an assistant who simply handles tasks that she does not have time to finish, such as grading assignments and summarizing key points from them. Though I still discussed what I did and showed my reflections to Priya, most work was practical and behind-the-scenes. By "behind-the-scenes," I mean my role was more like someone who contributed to the class in a less visible way, because I was just finishing all the tasks designed, organized, and evaluated by Priya. As time went by, I started to have my own understanding of the course material and the students. In later semesters, I started offering some of my own suggestions on the structure of the coursework. After Priya supported me in freely making suggestions on the course, I became more confident in terms of observing the class, leaving feedback, and sometimes even improvising a little bit when dealing with course-related affairs. Starting from the second semester, Priya encouraged me to give lectures based on one of my research projects, which is relevant to the class. She also discusses the structure, content, and activities of the class with me at the beginning of every semester. I transitioned from merely an assistant to a collaborator in the instructional team.*

Inspiration for Learning

Students enroll in degree programs for myriad reasons, not all of which center learning. For some, going to college is an expectation, while for others, a degree might represent a hard-fought ticket to a higher-paying job. Nevertheless, all students have the potential to learn and grow through their courses. Instructors can motivate students to learn by tailoring course material to student interests and connecting it to real-world issues (Ambrose et al., 2010).

After introductions on the first day of class, Priya provides an overview of the course. She begins with the observation that digital technologies have changed the way we communicate, socialize, work, play, and learn and that these changes pose significant social, cultural, economic, legal, ethical, and political questions. She then has students write a response to a reflection question. In the first few iterations, the question was straightforward: What do you hope to learn from this class? Since their answers were fairly generic, she now asks students: What fascinates you about digital technology? What worries you about digital technology? These questions narrow students' focus and allow them to bring a bit more of their own perspectives, which they then share during a brief class discussion. Students will revisit their answers on the last day of class and reflect on how their perceptions of technology may have changed over the semester.

After students have responded to the reflection questions, Priya shows recent news headlines pertaining to technology and highlights that the course covers technology-related issues that society is actively grappling with. She explains that while the course raises many questions, it provides few answers. Instead, the course gives students a framework to begin answering those questions in the present and to consider how they might continue answering them as they move into the workforce and adulthood. She then introduces what she hopes students take away from the course, which is to approach technology as a system, rather than a mere tool. This overview is meant to pique students' curiosity, inviting them to consider how the course material fits into their life, and to set the tone that they are welcome, and indeed encouraged, to contribute their perspective (Lang, 2019). Priya reinforces this throughout the semester by incorporating their work into the lecture materials. For instance, one of the early memo assignments asks students to create a six-word definition of technology, drawing on that week's readings. Before class, she clusters their definitions into patterns, and during class, she has students deduce the common characteristics. She highlights how the different definitions embody different worldviews and explains how the subsequent weeks' material will delve into them. She revisits the definitions throughout the course, to help students make connections between various course topics.

To demonstrate the real-world applicability of course topics, she draws on material published by Internet platforms themselves. For example, when discussing the right to be forgotten, a law that allows people to request that digital platforms delink search results from their name, Priya's lecture incorporates examples of actual requests that Google has received (which the company includes in its transparency reports). Students then guess the outcome and discuss. During the week that focuses on digital labor, particularly content moderation, students complete an in-class group activity that involves reviewing Facebook's community standards, brainstorming how an algorithm might detect specific categories of prohibited content, and discussing what aspects of the process may require human intervention. These activities use real-world materials to help students understand why inherently social processes—like those related to content removal—are so challenging to automate.

Reflection from Ananya: *The first few class sessions opened my eyes to concepts and social contexts surrounding technology that had never occurred to me before, even as a third-year Human-Centered Design and Development student. For example, the idea of a systems-based approach versus a tool-based approach to technology was completely new to me and fundamentally altered how I perceive technology by encouraging me to think with more nuance.*

Nuance is the main thing IST 431 left me with. I completed the class equipped with a multi-dimensional outlook on topics I once thought about in one dimension. For example, in the second half of the course, I was exposed to a report on four myths surrounding technology and well-being (Lenhart & Owens, 2020*). This reading challenged a lot of the hard-held beliefs that were instilled in me by popular media—that social media is inherently and inevitably bad. It opened my eyes to how we frame social media issues and often ignore the variety of experiences people have. The report highlighted how certain groups can benefit from the increased access and visibility that social media offers. To put everyone under the same umbrella by assuming that social media harms everyone equally can be a very narrow approach. The report also highlighted the power that we possess as users. Not only did this challenge an existing belief, but it was also one of the instances where the class reminded me that we have the power to steer technology to benefit us. "Human in the loop system" was the exact phrase I learned in another course reading about technosolutionism (Broussard,* 2019*), reminding me that prioritizing our power and expertise as humans is going to be important as technology progresses further.*

Another memorable course topic was dark patterns, especially because I was able to bring information I already knew from other classes into IST 431. As a student who studies user-friendly design, I encountered dark patterns a lot in my coursework, and I was excited to be able to draw on this prior knowledge. Dark patterns are deceptive tactics in user experiences usually designed to trick the user into doing something they don't realize they're doing, sometimes going against their best interest (Narayanan et al., 2020*). I signed up for a formal group presentation during the week on design because it included dark patterns. In many ways, the teaching team provided opportunities for students to bring their own strengths and skillsets to class; I distinctly remember Priya once asked a student for clarification on a topic she was presenting about. It's very emboldening and reminded me that we all have something to bring to the table, even in a university classroom where there is a clear expertise difference between the teaching team and the students.*

Reflection from Yao: *From our second semester working together, Priya gave me an opportunity to lead class sessions on surveillance. There are two reasons for this. First, I told Priya about my desire to be a professor in the future, and she therefore gave me the chance to practice my teaching skills. Second, Priya was familiar with my research interests and knew I had published papers on surveillance. Therefore, I took the chance and started to design the course materials and activities.*

The first thing I needed to do was to stay consistent with the style of the overall course. I referred to how Priya designed the classes in the previous semester and checked the template of the course slides. I also looked for material and designed activities to fit the template. For instance, I knew Priya often used short videos from social media or news media when introducing a topic to make the content more interactive. I also did the same thing.

However, this did not mean my experiences of leading class sessions were easy. I experienced a lot of difficulties. The difficulties were not about delivering the knowledge to the class, but about how to control the flow of the class. By flow I mean a state where students get engaged in the course material and the activities and then the class could proceed smoothly. The most difficult thing for me was to attract the students' attention and have them engage in the in-class discussion. Priya noticed this and shared some strategies with me. She gave me a bag of cards with students' names. If nobody responded to a discussion question, I could pick one card from the bag and ask the student to answer. This helped me a lot.

I also learned a lot through my own experiences and reflections. The first one is the mindset to be an instructor. In the first classes, I was quick to panic if I encountered something unexpected, such as no response from students or spending too much time on one part and leaving little time for the next part. My brain stopped working for a couple of seconds. I felt embarrassed and just wanted to leave the class. After a couple of classes, I felt calmer. Additionally, I also tried some techniques and kept using them if they worked. For instance, when a student mentioned a point in the class, I would refer to the student if someone in the later discussion also mentioned a similar one. This also helped me make the class smoother and more engaging.

A Web of Significant Relationships

While the instructor is administratively and intellectually in charge of a course, it is neither practical nor desirable for the instructor to be the source of all learning and support. Students value opportunities to learn from their peers and may even learn more effectively from them. Beyond the classroom, colleges and universities offer a plethora of learning activities and support services, but students may need reminders about their existence as well as encouragement to take them up (Felten & Lambert, 2020).

Given the importance of active and peer learning, course activities include a mix of paired and group work. Lecture-based class sessions include think-pair-share activities, where students reflect on a question related to the day's topic, discuss their thoughts with the person sitting next to them, and then share their reactions with the whole class. Group formation for in-class activities varies between clustering students who sit near each other and random assignment. Since students often sit in the same places, they end up primarily working with the same people over the course of the semester. As the semester progresses, small talk and banter arise during interstitial moments in class, suggesting a positive level of rapport and comfort among at least some students. Random group assignments enable them to continue getting to know other classmates throughout the semester.

While students informally present at the end of each in-class activity, there is one formal group presentation assignment. As noted in the course structure section, Tuesday class sessions begin with a student presentation introducing the week's topic. Students sign up for a week at the beginning of the semester, and they are instructed to meet as a group beforehand to coordinate their presentation. They conclude the presentation by posing a discussion question to the class. Students then write their responses in the learning management system, and brief discussion ensues.

Throughout the course, Priya informs students of the myriad opportunities for engagement, connection, and support available across campus. She verbally reviews course policies during Week 1, reminding students to work with campus disability services to receive accommodations and encouraging them to make use of the university's various mental health and well-being services. Over the semester, Priya shares announcements about campus events related to course content, as well as college and university-wide opportunities such as career fairs or scholarship competitions. This information typically reaches Priya through campus listserv emails, which often contain text or slides that she can easily insert into course materials. Class slides always include links where students can find more information. Periodically, Priya invites guest speakers to present on topics related to the course, for instance, a journalist who published a book on political misinformation.

In Week 3, when students are developing their final paper roadmaps, the IST librarian comes to class and walks through the research process, reviewing the various library resources available for students. The librarian then leads an activity where students locate and share potentially relevant sources to use in their research. The librarian also shares a library guide she created specifically for the final paper assignment and invites students to meet with her as they develop their papers.

Reflection from Ananya: *A defining factor of IST 431 was its focus on collaboration and interaction, both with peers and with the teaching team. It is one of the few classes where I still remember several of my classmates. I recognized many of my IST 431 peers outside of the classroom. I remember one instance where I was walking with one of my classmates and we continued a discussion from class that day. While it is common to make friends who take the same class as you, I think it is less common to connect over course topics outside of class.*

I was also able to connect with Priya over research, incited by the slide she put up during class inviting everyone to her research presentation. I reached out to her about this, which eventually led to my position as a research assistant under her supervision. This was an outcome I still greatly cherish. It has enabled me to develop a research study alongside her, stemming from my own research interests. As an undergraduate, I am very grateful that I had such an opportunity in the first place, where my ideas and interests were molded into actual academic artifacts. I was able to learn an immense amount from Priya about the world of academia and the ins and outs of conducting informatics research.

Reflection from Yao: *Because of my primary role (grading and evaluating students' performances) in the class, I paid less attention to social networking with students to avoid bias. But I did expand my network because of this course. After collaborating with Priya for one semester, I found we have similar research interests, and I admired her work style. So, I invited her to be a member of my PhD dissertation committee. Since then, I have learned a lot from her not just about how to be a teacher but also about how to be a researcher.*

Reflection from Priya: *My ability to cultivate a relationship-centered learning environment stems directly from the fact that the institution has committed resources* *toward instructional support. I joined the IST faculty with two semesters of experience as a graduate teaching assistant, and when I started, I had two weeks to prepare to teach my first class. Institutional support was an anchor amidst the flux. Two senior faculty who had taught IST 431 generously shared their materials with me. The college's Office of Teaching, Learning, and Assessment paired all new faculty with an instructional designer and organized a series of webinars before and after the semester started to orient us to the learning management system and the basics of course design. The university's Teaching and Learning with Technology office offered a webinar series for new faculty, and the Schreyer Institute for Teaching Excellence organized a year-long new faculty group meeting series. Through these efforts, I quickly established relationships with staff across campus whose expertise was teaching, equipping me to find my bearings as an instructor. Now, several semesters into my time at IST, the people I met through these efforts are some of my most trusted colleagues.*

Exploration of Meaning and Purpose

Teaching in IST means having the good fortune of covering topics that can easily connect to students' lives, current events, and the wider world. But since many IST courses focus on helping students develop specific skillsets or knowledge bases, it's important for courses like IST 431 to give students space to reflect on the role that technology plays in society and learn how to question its often rapid embrace. Whether grappling with their own complex relationships with social media, wondering how exactly algorithms fuel misinformation and polarization, or welcoming the development of AI to streamline work processes, students are already thinking about the interplay between technology and society. They deserve room to explore those questions in relation with peers and mentors (Felten & Lambert, 2020).

Since the goal of IST 431 is to shift how students think about technology, their final summative assignment involves writing a paper that investigates a technology-related question that interests them. To help structure the research process, the assignment explains that the paper should provide background information and context so the reader understands where the guiding question is coming from, identify the stakeholders affected by the guiding question, answer the question, explain the social consequences of the question so the reader understands why the question matters, and offer recommendations for next steps that can be taken to further address the issues implicated by the question.

As described in the previous section, students complete an in-class activity in Week 3 to help them develop and refine their guiding question. During class, Priya and Yao walk around and provide in-the-moment feedback. After this and the aforementioned activity with the IST librarian, students submit a roadmap outlining their ideas in response to the final paper prompts, which Priya reviews. In Week 8, students submit a first draft of five pages. They then exchange drafts with a partner and conduct a peer review. Priya assigns partners in advance, usually pairing students with similar topics. Priya also reviews each student's first draft, providing comments and advice. During class, she summarizes her guidance into recommendations for all students to consider as they continue to work on their papers. She consistently tells students to use the feedback to strengthen their work and encourages them to meet with her if they have any questions or are feeling stuck, which several students have done.

While Priya has always encouraged students to use the final paper to explore a question they find personally meaningful, she recently incorporated reflection into the assignment instructions. A few weeks into the fourth iteration of the course, she encountered the concept of "I-Search" papers in a teaching newsletter (McMurtrie, 2023). These are papers where students develop and explore a research question, but their write-up includes more explanation about their process and reflection on what they learned through it. That day in class, she asked students to provide anonymous feedback on whether an I-Search paper interested them. Several said yes and a few said no. So, she incorporated I-Search-related prompts into the paper instructions but made them optional.

Recognizing that not all students feel comfortable with writing and that the rise of tools like ChatGPT may tempt students to take shortcuts in writing assignments, Priya seeks to introduce more flexibility into the way students present the work they put into exploring their guiding question. The fact that some students preferred *not* to do an I-search paper serves as a reminder that care needs to be taken when incorporating personal reflection into assignments so that they scaffold metacognition without requiring students to disclose more than they may feel comfortable sharing. (As a privacy scholar, Priya is particularly attuned to questions of disclosure.)

Reflection from Ananya: *Out of all the assignments, I was most excited about the final paper because it gave us free rein to select a topic of our choosing. I enjoy writing and find it to be one of my stronger skills, so I was looking forward to the challenge of drafting a full, nuanced, research paper. I was also working on the final paper in tandem with my work as a research assistant for Priya, so there was some interesting mental overlap that occurred, when I would apply some of my research knowledge from my work position to my coursework.*

*My guiding question for the final paper was "How does social media design affect adolescent attention spans?" I knew early on that I wanted to write about attention. This is mainly because I kept reading and hearing about attentional issues and things like "dopamine detoxes" to supposedly "fix" our attention. As a child of the Internet age, I have been around technology for as long as I can remember, and I felt that the topic directly pertained to me in several ways. I had found that I had a much shorter attention span compared to when I was in my early teens, and I wondered how my social media use may have contributed to it. When I first came up with the question midway through the semester, I had a pretty definite idea of my own stance on the topic. However, as we entered the second half of the course and I read the Lenhart and Owens (*2020*) report, alongside others, I realized that my paper could do with a lot more nuance than it had in the first draft.*

Overall, the experience of writing the paper taught me a lot. It was one of the experiences that cemented my interest in academic research, which is important for me to know about myself as I approach graduation. While I had written several course papers before, this paper reflected some of my most nuanced arguments and ideas in writing. It was a reflection of 3 years of college and a semester-long learning experience that taught me how to be a better thinker in my field.

Conclusion

As of Spring 2024, Priya and Yao have worked together for 2.5 years, and Ananya and Priya for 1.5 years. We recognize that a lot about our situation is unique: instructors and teaching assistants may have little control over what they teach or who they work with, and most undergraduates will not go on to lead research studies with faculty. However, building relationships in the classroom is just as important as striving toward learning outcomes. Indeed, the two are connected, as students benefit academically and personally when they make meaningful connections with their instructors (Hagenauer & Volet, 2014; Mayhew et al., 2016; Schwartz, 2019; Felten & Lambert, 2020). And for many instructors, forging those connections is the most rewarding aspect of academic life. In this chapter, we have illustrated how an upper-level undergraduate course on technology and society embodies the principles of relationship-rich education (Felten & Lambert, 2020) and reflected on how such course design has affected the experience of those involved. We conclude with brief observations for instructional teams and administrators involved in teaching-related decision-making.

For instructional teams, we believe that the most important element of relationship-rich education is to approach teaching with a mindset of partnership. While students are ultimately responsible for doing the hard work of learning, the instructional team plays a significant role in setting students up for success. Course policies, assignments, and assessment processes signal to students how much instructors trust them and believe in their potential, and we encourage instructional teams to design courses that approach students as learners, not adversaries (Denial, 2019). Throughout this chapter, we have offered examples of how this manifests in IST 431. But centering relationships is not simply a matter of course design; the attitude instructors bring toward teaching also matters. We highly recommend that anyone involved in teaching read Denial's (2019) essay on a pedagogy of kindness for a personal account of how one professor shifted away from the student-as-adversary mindset after more than two decades of teaching. We hope that our personal reflections throughout this chapter demonstrate the value of centering relationships in teaching.

For administrators involved in teaching-related decision-making, we emphasize that relationship-centered education is not something instructional teams can foster on their own. Relationships are more likely to thrive when instructional teams work in an environment that supports them, especially instructors early in their careers.

Policies on teaching loads, the allocation of teaching assistants, and the availability of instructional support all shape how and how much instructors feel supported. At minimum, departments and colleges should tell new faculty and teaching assistants what instructional support resources exist on campus and connect them with a specific person or team who can help them acclimate to their educational role. Dedicated events, from webinars to workshops to open-ended discussion sessions, give instructors the time and space to connect with others while honing their skills. The more a university invests in cultivating a culture of support, the more easily instructors, teaching assistants, and students can forge the meaningful connections that drive learning and personal growth.

References

Ahad-Legardy, B., & Ocampo, A. (2021). When did you fall in love with reading? (No. 6) In *Professor-ing*. National Center for Faculty Diversity and Development. https://podcasts.apple.com/us/podcast/when-did-you-fall-in-love-with-reading/id1519285794?i=1000511373948

Ambrose, S. A., Bridges, M. W., DiPietro, M., et al. (2010). *How learning works: Seven research-based principles for smart teaching* (1st ed.). Jossey-Bass, a Wiley Imprint.

Broussard, M. (2019). Letting go of technochauvinism. *Public Books*. https://www.publicbooks.org/letting-go-of-technochauvinism/. Accessed 29 Jan 2024.

Denial, C. (2019). A pedagogy of kindness. *Hybrid Pedagogy*. https://hybridpedagogy.org/pedagogy-of-kindness/. Accessed 29 Jan 2024.

Felten, P., & Lambert, L. M. (2020). *Relationship-rich education: How human connections drive success in college*. Johns Hopkins University Press.

Hagenauer, G., & Volet, S. E. (2014). Teacher–student relationship at university: An important yet under-researched field. *Oxford Review of Education, 40*, 370–388. https://doi.org/10.1080/03054985.2014.921613

Lang, J. M. (2019). How to teach a good first day of class. *The Chronicle of Higher Education*.

Lenhart, A., & Owens, K. (2020). *Good intentions, bad inventions*. Data & Society.

Mayhew, M. J., Rockenbach, A. N., Bowman, N. A., et al. (2016). *21st century evidence that higher education works* (2nd ed.). Jossey-Bass.

McMurtrie, B. (2023). Rethinking research papers, and other responses to ChatGPT. *The Chronicle of Higher Education*.

Narayanan, A., Mathur, A., Chetty, M., & Kshirsagar, M. (2020). Dark patterns: Past, present, and future. *Communications of the ACM, 63*, 42–47. https://doi.org/10.1145/3397884

Schoem, D. (2017). Relational teaching and learning. In D. Schoem, C. Modey, & E. P. St. John (Eds.), *Teaching the whole student: Engaged learning with heart, mind and spirit* (1st ed., pp. 79–99). Stylus Publishing.

Schwartz, H. L. (2019). *Connected teaching: Relationship, power, and mattering in higher education* (1st ed.). Stylus Publishing.

Achieving Consistent and Relevant Learning Outcomes on a Common Course

Rosalie J. Ocker and Lisa Lenze

Abstract This chapter describes efforts to revitalize a foundational course in database design and development. Required by most majors in the College of IST, numerous course sections are offered each semester. The College of IST is composed of multidisciplinary faculty across domains including computer science, human-computer interaction, information systems, and privacy and security. This diversity of faculty academic backgrounds is beneficial to the college for many reasons. However, it can be an impediment to offering a cohesive foundation course where all students need to be proficient in a set of core concepts. We discuss the process to move toward a shared understanding by faculty of essential course components, the development and deployment of a specified database course, where course content is provided to faculty in a ready-to-teach format, student results assessing achievement of learning objectives, and faculty feedback on course effectiveness. The goal of these efforts is to provide faculty with an out-of-the-box course that meets learning objectives, is ready to use, and is embraced by faculty and students across the various majors in the college and the campus locations where it is taught.

Keywords Introductory Database Course · SQL · Teaching consistency · Course standardization · Active learning

Introduction

The fundamentals of database (FDB) course stands as a foundational requirement in five out of six majors within the College of Information Sciences and Technology (IST). Delivered not only at the University Park (main) campus but also at various regional campuses across Pennsylvania and via PSU's World Campus (an online

---

R. J. Ocker (✉) · L. Lenze
The Pennsylvania State University, University Park, PA, USA
e-mail: rxo4@psu.edu; lfl2@psu.edu

J. M. Carroll (ed.), *Innovative Practices in Teaching Information Sciences and Technology*, https://doi.org/10.1007/978-3-031-61290-9_18

platform), this second-year course serves a diverse student body each semester. Many regional campuses follow a 2 + 2 model, where students spend their initial 2 years of study at regional locations before transitioning to University Park campus for the remaining duration of their degree.

The FDB course lays the essential groundwork for advanced studies. However, its impact has been hindered by inconsistent content and teaching methods. The course content has been found to significantly depend on the individual instructor leading a specific section, resulting in disparities across course sections. For instance, while some instructors prioritize database design, crucial elements such as mastering structured query language (SQL) may be a minimal part of the content, despite its significance in advanced coursework.

Thus, this lack of standardized instruction poses challenges in upper-level courses, where students may not possess the requisite knowledge base. Recognizing this disparity, efforts were undertaken to specify and standardize course content across FDB course offerings. Addressing these inconsistencies is vital in ensuring a uniform learning experience, thereby equipping students with a robust foundation for their future studies and careers. This paper delves into the initiatives taken to establish consistent and relevant learning outcomes in the FDB course.

To better understand the challenges faced in achieving a consistent and relevant learning outcome, we first address the interplay of organizational structure and distance in relation to achieving a shared understanding of course content and delivery. Second, we discuss pertinent pedagogical approaches and critical instructional elements that promote attainment of learning goals.

Organizational Structure of Penn State

The Pennsylvania State University (PSU) is a large university with campuses spread across the state. Ninety-six percent of Pennsylvania residents live within thirty miles of a Penn State campus. In undergraduate education, it includes University Park campus in the center of the state and 19 regional campuses across the state (see Fig. 1). University Park (UP) campus along with the regional hubs offers face-to-face courses, ensuring a traditional learning experience for students. Additionally, PSU extends its educational reach through the World Campus, providing online degree programs exclusively. Beyond geographic distinctions, each regional campus boasts its own faculty, leadership structure, and collaborative initiatives. These campuses often host independent sports teams and extracurricular activities, fostering a unique identity within their local communities. Notably, 14 regional campuses have unified to create university college, which functions as a cohesive unit to advance shared strategic initiatives.

The UP campus overshadows regional campuses in terms of size of the student body and faculty. Undergraduate enrollment is different across campuses; UP campus (45,000) is four times the size of university college and twice the size of World Campus.

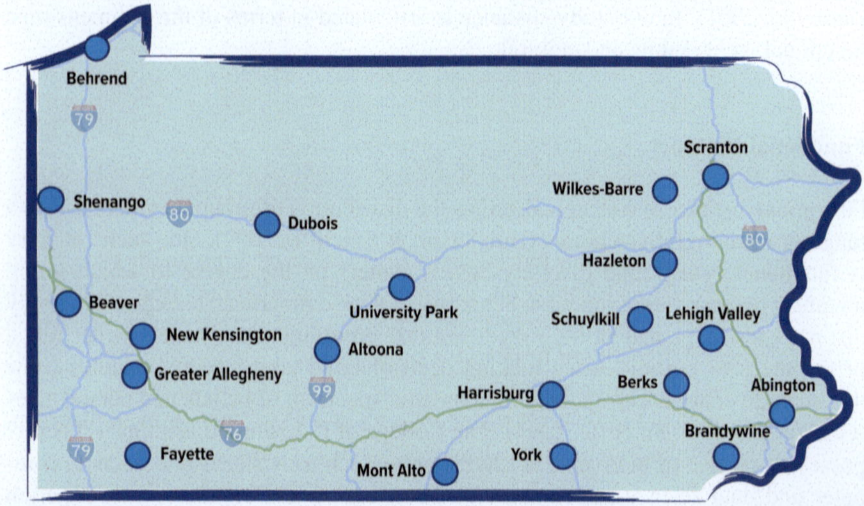

Fig. 1 Map of Pennsylvania showing campuses with undergraduate students

In addition to UP campus and World Campus, the FDB is taught across 15 regional campuses. Dozens of instructors teach FDB every year.

Shared Understanding

Shared understanding is critical to achieving desired outcomes (Jenkin et al., 2019). Shared understanding denotes the extent to which stakeholders collectively grasp a particular subject or approach, encompassing aspects like goals and desired outcomes (Gregory et al., 2013). Shared understanding delves into the alignment of perspectives, where individuals agree on the value of properties, the interpretation of concepts, and the mental models explaining cause and effect within a given context (Bittner & Leimeister, 2014).

The central premise of shared understanding posits that the effectiveness of any group significantly improves when its members share a comprehensive understanding of the task, group dynamics, equipment, and the overall situation (Mohammed & Dumville, 2001). This shared comprehension serves as a bedrock for collaboration, ensuring that all involved parties are on the same page, thereby enhancing overall group performance and outcomes.

Distance

Distance impedes shared understanding (Ocker et al., 2011). Various types of distance are described in the literature (Espinosa et al., 2003, Mohammed &

Dumville, 2001). In our study, distance is articulated in terms of three dimensions: functional, geographic, and cultural.

Functional Distance

Functional distance is conceptualized as the distribution of group members across a range of meaningful categories (Bunderson & Sutcliffe, 2002). One such category is functional background diversity, which centers on the degree to which group members possess varied functional expertise. Diverse functional backgrounds imply distinct knowledge and skills, suggesting that group members have a wider range of resources to tap into when making decisions and taking action. Studies show that such diversity is correlated with a broader spectrum of beliefs and perspectives (Glick et al., 1993; Sutcliffe, 1994). The College of IST is interdisciplinary. Faculty possess expertise from diverse academic areas such as artificial intelligence, computer and data sciences, cybersecurity, human-computer interaction, information systems, privacy and security, and the social sciences. These different backgrounds bring a diversity of approaches and viewpoints which enrich the college in many ways.

However, in relation to the FDB course, the diverse academic backgrounds of faculty lead to disparate approaches to teaching the course. For instance, some faculty members prioritize database design, dedicating limited time to learning structured query language (SQL). SQL stands as the predominant coding language for extracting and organizing data stored in relational databases. Mastering SQL is crucial for students enrolled in advanced database courses and is equally vital for those entering the job market, where the ability to extract data from relational databases is a fundamental skill needed by many students graduating from the College of IST.

Geographic Distance

Many groups encompass members that are collocated with some members that are geographically distant. Members within a given location are physically collocated–they interact primarily in a face-to-face manner. They share the same context and as such are affected by the same organizational culture, work processes, and resources. The frequent face-to-face contact of collocated members increases the likelihood that they share an identity (Hinds & Mortensen, 2005).

Geographic distance separates all campuses, as depicted in Fig. 1. Regional campuses are physically separate from one another and University Park. This distance impedes collaboration and coordination across campuses, where initiatives are often seen through an "us vs. them" lens. While many regional campuses joined together to form university college, this added structural layer serves to increase shared identity among these regional campuses, but which increases separation from University Park.

Cultural Distance

From a broader perspective, culture pertains to the home country of individuals. However, on a more specific level, the organization or subunit within which group members operate also possesses its unique culture. In situations where groups are dispersed within the same organization, the organizational culture significantly shapes interactions and group dynamics. Martins (2003) classifies organizational culture into three perspectives: integrated, differentiated, or fragmented. An integrated culture signifies a strong consensus on values, assumptions, and behaviors, fostering unity and a shared identity among organizational members, guiding their attitudes and actions. Resistance to new initiatives can stem from their misalignment with the values ingrained in a robust organizational culture.

In a differentiated culture, subcultures exist, where a homogeneous culture prevails within each group. Conflicts between stakeholders arise due to incompatible interests within these subcultures. Conversely, a fragmented culture is characterized by pervasive contradiction. In such cases, the culture embodies ambiguous and contradictory symbols and behaviors, leading to intricate dynamics and challenges within the organization.

While the university slogan "We Are. . . Penn State" is embraced across all campuses and attests to an integrated organizational culture at a macrolevel, differentiated cultures exist across campuses. Additionally, regional campuses that are part of university college share leadership and operate as a bloc on strategic and operational issues. Thus, these campuses are bound together through organizational structure and shared governance. Additionally, World Campus, launched in 1998, is a well-established entity within PSU. It caters to adult learners, many of whom are part-time students located across the globe. Table 1 summarizes key points about distance impact on the development of shared understanding.

Table 1 Impact of distance on shared understanding at PSU

Types of distance	Embodiment at PSU	Impact on shared understanding
Functional distance—functional background diversity	Different academic backgrounds lead to different approaches to teaching the FDB course	Reduces shared understanding regarding course content and teaching approach
Geographic distance—physical distance between campuses	Faculty at a given campus have a shared context. Their identity is distinct from University Park campus	Reduces shared understanding due to lack of shared context across campuses
Cultural distance—organizational culture	Campuses have distinct organizational cultures and governance structures	Reduces shared understanding due to different subcultures

Pedagogical Approach

To promote students' attainment of the FDB course learning goals, the primary author designed the course congruent with meaningful learning theory, active learning principles, goal-directed practice, and targeted feedback. All instructional materials support this pedagogical approach.

Meaningful Learning Theory

Meaningful learning theory (Ausubel, 1963), a constructivist epistemology, describes learning as the process of making sense of new information by integrating learners' prior knowledge with newly introduced concepts, skills, or procedures and then applying the newly formulated knowledge to new situations. Key elements in meaningful learning theory are prior knowledge and active construction of new knowledge.

The paradox of building new knowledge on prior knowledge is that prior knowledge is often woefully deficient and yet, at the same time, the only foundation that the learner possesses for constructing new understandings (Roschelle, 1995). These often-faulty foundational structures, referred to as limited or inappropriate propositional hierarchies, require conceptual change (Novak, 2002). Critical instructional elements that promote conceptual change and meaningful learning include opportunities for active learning, goal-directed practice, and targeted feedback.

Active Learning

Active learning, a term dating back to the 1980s and 1990s (Cross, 1987; Chickering & Gamson, 1987; Bonwell & Eison, 1991), refers to teaching pedagogies and instructional activities aimed at engaging students in constructing meaning, typically during an in-person class period. Active learning includes both collaborative pedagogies such as problem-based learning (Allen et al., 2011; Hung et al., 2008), team-based learning (Michaelsen & Sweet, 2008), cooperative learning (Slavin, 1996), and peer instruction (Fagen et al., 2002; Crouch & Mazur, 2001) and individual and paired learning activities such as think-pair-share (Kaddoura, 2013; Kothiyal et al., 2014), flipped classrooms (Day & Foley, 2006; Foertsch et al., 2002; Wilson, 2013), and hands-on labs and exercises (Malik & Zhu, 2023).

The central premise of active learning asserts that students must actively engage with the subject matter in order to develop rich understandings. This engagement may include problem-solving, analysis, synthesis, evaluation, or other higher-order activities (Bonwell & Eison, 1991) in written or oral modes. Active learning pedagogies and instructional activities facilitate meaningful learning outcomes such as accurate conceptual knowledge, retention of knowledge, transfer of knowledge to new contexts, and robust problem-solving skills (Crouch & Mazur, 2001; Fagen et al., 2002; Hake, 1998; Halpern & Hakel, 2003; Weiman, 2014). STEM students who engage in active learning courses are less likely than their peers in lecture courses to fail the course, and they score higher on exams and concept inventories (Freeman et al., 2014).

Students who participate in active learning activities practice what they are learning–whether that is through problem-solving, critical thinking, refinement of understandings, retrieval of information, or other activities; and they typically receive feedback at the conclusion of the activity. What makes that practice and feedback most instrumental to learning?

Goal-Directed Practice

Adages such as "practice makes perfect" and "use it or lose it" express the importance of retrieving information over and over–a key factor in learning (Halpern & Hakel, 2003; Oakley, 2014). However, time on task, alone, does not lead to meaningful learning. Hattie (2009) explains "we certainly do not want more time on task if the learning is not positive" (2009, 184). In contrast, goal-directed practice leads to meaningful learning. It has characteristics of being deliberate, appropriately pitched at the right level, and repetitive, with opportunities for successive refinement (Hattie, 2009).

Ambrose et al. describe goal-directed practice with the same three elements—"practice that (a) focuses on a specific goal or criterion for performance, (b) targets an appropriate level of challenge relative to students' current performance, and (c) is of sufficient quantity and frequency to meet the performance criteria" (2010, 127). This type of directed practice is the sort of practice undertaken by accomplished musicians, professional athletes, master chess players, excellent typists, and experts in other fields (Anders Ericsson, 2008).

In our work, these three elements defined the structure of the activities created to teach students SQL. Students encountered new concepts with pre-class interactive readings. These pre-class activities define the goal. A series of follow-on assignments provide practice opportunities with incremental levels of difficulty and supported incremental independence. During in-class exercises for resident instruction, students work on problems with the instructor and peers. Students work on slightly more complicated problems in lab exercises that they begin during class and finish independently after class.

Targeted Feedback

In addition to ample and well-designed opportunities for practice, learners require targeted feedback (Anders Ericsson, 2008; Halpern & Hakel, 2003). In Hattie's (1992) review of literature on influences on learning, he concludes from his analysis of 134 meta-analyses that feedback was among the most powerful influences.

A critical element of targeted feedback is goal-based communication. Goal-based communication begins with sharing clear goals for students' learning and culminates in providing (a) an account of students' progress and (b) direction on where to focus attention for subsequent effort (Ambrose et al., 2010; Hattie, 2009). Feedback with these characteristics is more effective than providing generic feedback (Black & William, 1998).

Targeted feedback can be communicated through individually written comments, computer-generated responses, trained peer evaluators, and carefully written analytic rubrics that express the criteria for meeting assignment goals (Hattie, 2009). Analytic rubrics, when shared ahead of time, focus students on the target and, when completed, show students what they achieved and where they need to focus. As such, they can promote learning. Jonsson and Svingby explain that "the main reason for this potential lies in the fact that rubrics make expectations and criteria explicit, which also facilitates feedback and self-assessment" (2007, 130).

Goal-based communication falls short if it is not delivered in an appropriately timed manner. Ambrose et al. explain the timing of feedback with a GPS analogy: "it is clear that one of the features of these devices is that they give feedback when the driver needs it to support the goal of reaching a particular destination" (2010, 142). Appropriate timing of feedback depends on the learning goal (Ambrose et al., 2010). In introductory courses such as the one described in this chapter, instructors set goals to learn a particular concept, skill, or procedure; and quick feedback is most useful. Leinonen et al. (2022) study of feedback on computer science course projects highlights the benefit of setting early milestones for projects, so that students receive feedback along the way (2022, 890).

In our work, students receive quick, direct, and staged feedback, tied to rubrics that express the criteria for the assignment. During in-class exercises, students' SQL code either works and they get the answer, or it does not work, and the instructional team helps students on the spot. For practice exercises which are begun in class and completed outside of class, instructors review the exercises and answer questions in class after students turn in their work. Students may submit their work again and the graders grade only the latter submission. In this way, students spend time on the practice exercises twice, with individual feedback during the class period when the exercise is assigned and en masse feedback before submitting the assignment for evaluation. Table 2 summarizes the core concepts and principles related to meaningful learning theory.

Goal of Course Redesign

Given the organizational and distance constraints, and considering the tenants of meaningful learning theory, the goal of the FDB course redesign is threefold. Firstly, it must foster a shared understanding of the course content among faculty. Secondly, it must encompass a standardized course structure, ensuring that all sections of the course cover the same content using the same materials. This uniformity

Table 2 Core concepts and principles of meaningful learning theory

Concept	Description
Meaningful learning theory	Learning is the process of integrating prior knowledge with new information for better retention and application. Active engagement, goal-directed practice, and targeted feedback are critical for meaningful learning
Active learning	Pedagogies and activities engaging students in constructing meaning, including problem-based learning, cooperative learning, and hands-on exercises. Active learning promotes accurate conceptual knowledge and problem-solving skills
Goal-directed practice	Deliberate, appropriately challenging, and repetitive practice focused on specific goals. Contrasts with mere time on task, emphasizing the quality of practice for meaningful learning
Targeted feedback	Crucial for learning, feedback should be goal based, providing clear goals, progress assessment, and guidance for improvement. Timely feedback is essential, and various methods like rubrics or computer-generated responses can be effective

across sections enhances consistency and quality of student learning. Lastly, the redesign must emphasize an interactive learning approach, encouraging active student participation around a set of goal-directed activities that incorporate targeted feedback. This hands-on learning environment promotes engagement, requiring students to interact with course content, to foster a deeper understanding of the material.

Process of FDB Course Redesign

The method to establish a consistent approach to FDB course content and delivery is described in this section. The process of developing the course revision was undertaken in several distinct phases over a span of several years.

In the initial phase, which spanned 18 months, faculty teaching the course were invited to address the issues with the course. Faculty engaged in several in-depth and sometimes heated discussions regarding course learning objectives. These meetings were conducted on the University Park campus and included faculty who regularly taught the course at UP and World Campus. The discussions resulted in an updated document of modestly refined learning objectives.

Two UP faculty created a 20-question multiple choice quiz to assess the SQL portion of the learning objectives. The SQL quiz, referred to as the "common quiz," was given across multiple sections of courses at UP to gauge student learning prior to the course redesign. Quiz results confirmed that students differed on SQL knowledge by course section, indicative of the different emphasis placed on varying course learning objectives. The quiz questions underwent factor analysis and refinements were made to enhance reliability of the quiz as a measurement instrument.

The subsequent 3-month planning phase involved a review of the existing course template used by some World Campus faculty. During this phase, an interactive zyBook textbook utilizing MySQL, a widely used relational database by Oracle, was identified by the first author and adopted. MySQL Workbench is a graphical tool for working with MySQL servers and databases.

During the 3-month course redesign and development phase, the course was designed by the first author with assistance from one graduate student, one undergraduate student, and one learning designer. Over 40 interactive, hands-on exercises were created to facilitate active learning. The exercises were complemented by a series of labs, designed to reinforce key concepts. An end-of-semester project was also developed to tie together course content, where students would first design and then build a relational database using MySQL.

The pilot study of the redesigned FDB course was conducted across two 15-week semesters by the first author. The SQL multiple-choice common quiz was included as part of the final exam. Outcomes were reviewed and student feedback was obtained from the end-of-semester course evaluation. Minor revisions were made by the first author and the pilot study continued in the following semester with two additional sections of the FDB course taught by the first author.

Moving into the 3-month revision phase after the pilot studies, existing course exercises were revised, and new exercises were added. The revised course redesign consisted of over 40 in-class exercises spread across six two-week sprints. Each sprint included a lab and a unit quiz. The course culminated in a 3-week group project and a final exam.

Finally, the course entered the beta test phase. The "out-of-the-box" version of the FDB course was used by a group of faculty members who volunteered. The beta faculty group consisted of four faculty at UP and two faculty at World Campus. Faculty met as a group to share experiences and help each other during and after the 15-week fall semester. Some modifications were made between semesters, by two participating faculty at UP (not including the first author). The beta test continued into the spring semester. Only minor adjustments were needed at the end of the spring semester.

The following fall semester, the FDB course entered the rollout phase. All faculty assigned to teach the FDB course at UP and World Campus use the redesigned, "out-of-the-box" course materials. This course is now "specified," meaning that the redesigned course content, layout, materials, quizzes, and exams are provided to faculty with the expectation that they are used as provided.

Once each semester (twice per year), faculty teaching in IST courses from regional campuses and World Campus have a statewide meeting. During these meetings, the first author presented updates on the FDB revision. These presentations served to build awareness and intrigue among regional faculty without using a hard sell tactic. In this way, regional faculty were brought into the discussion of the FDB revision and invited to participate.

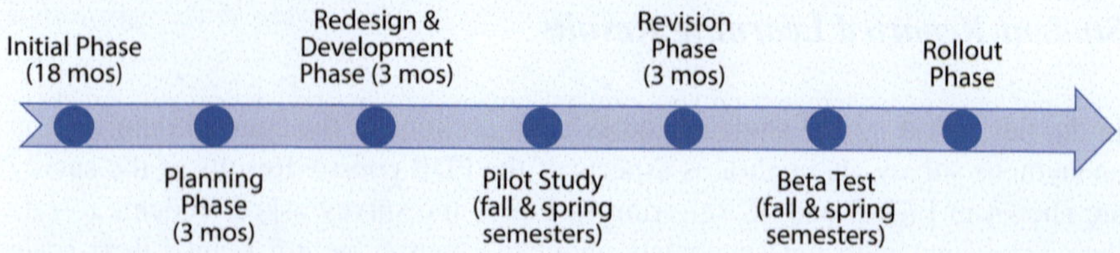

Fig. 2 FDB Redesign course timeline

This deliberate and iterative approach, spanning several years and integrating feedback from various stakeholders throughout the process, helped to ensure the development of a robust course. The timeline of the redesign efforts is shown in Fig. 2. Note that elapsed time from the initial phase to the course rollout was 4 years.

Findings

During and after the beta test, data were collected from students and faculty to ascertain the effectiveness of the redesigned FDB course.

Common Quiz Results

Seven of the eight sections offered at UP in the fall beta test participated. (One UP faculty member did not "buy into" the course redesign and taught the course as they did normally.) Additionally, two sections offered by the World Campus participated. In total, over 300 students participated across nine course sections taught by six different faculty members in the fall semester beta test.

The common quiz (20 questions) was again administered as an objective measure to ascertain student knowledge of SQL. The results of the common quiz indicate that the redesigned course was successful in meeting the FDB learning objectives. As shown in Table 3, the average score was above 75% for each beta course section, ranging from a low of 76% to a high of 87%; two-thirds of the sections scored an average of 80% or higher.

Table 3 Average common exam scores across beta sections

Section	A	B	C	D	E	F	G	H	I
Score	81	80	83	77	87	84	76	87	77

Student-Reported Learning Results

In the fall beta test, 229 students across eight sections of the course completed an anonymous survey about various aspects of the FDB course. Results of the survey are shown in Figs. 3 and 4. Referring to Fig. 3, the survey asked students to rate their agreement with ten statements about the course, on a 7-point Likert scale from strongly disagree to strongly agree. The results were positive. Over 80% of students agreed or strongly agreed that the instructional materials increased their knowledge and skills, that expectations for student learning were clearly defined, and what they were learning was important. About 70% of students agreed or strongly agreed that the various aspects of the course (zyBook, exercises, labs, and lectures) complemented each other and that the readings and organization of course content aided their learning.

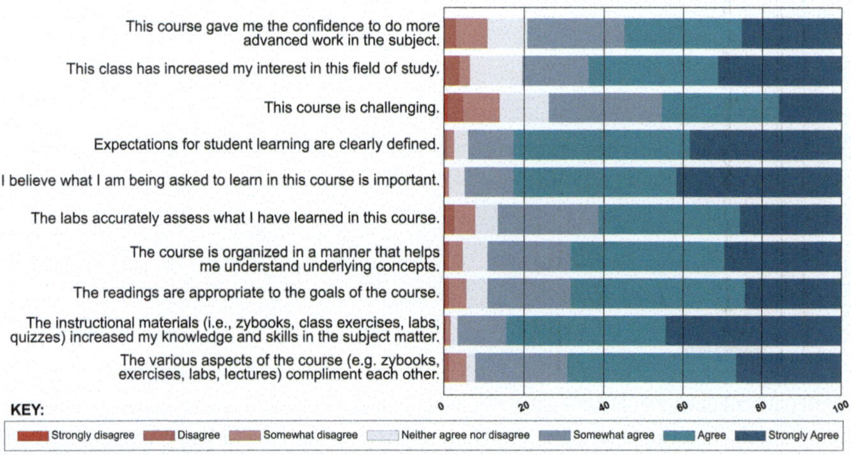

Fig. 3 Summary of student responses to an anonymous survey

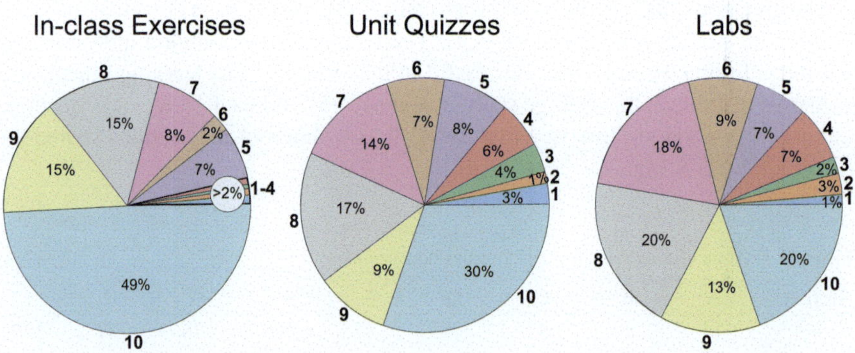

Fig. 4 Student ratings on keys aspects of the course

Figure 4 shows three pie charts that summarize student ratings on key aspects of the course, on a scale from 1 (low) to 10 (high). Students rated the in-class exercises as highly beneficial to their learning, with nearly 50% giving the exercises the highest rating and nearly 80% rating it at 8 or higher. Interestingly, students rated the unit quizzes (i.e., tests) as important to their learning, with over 50% rating the exercises 8 or above. Similarly, over 50% of students rated the labs, which build on the content of the exercises, at 8 or above. These aspects account for the core content of the course.

Faculty Feedback

In two semesters, we systematically sought feedback from faculty teaching the revised FDB course. The first feedback collection occurred after the fall beta test was completed, and the second occurred in the fall semester, 1 year later, after the revised beta version was rolled out.

Fall Beta Feedback Session

Three instructors who taught the revised course in the fall beta test participated in a faculty feedback session to debrief their experience and provide suggestions for the next course iteration. The first author, who authored the beta version and taught sections of the course, attended the session; her role was to listen and to assign requested changes to improve the course to the appropriate person. All four instructors taught the course in residence at UP.

We asked three questions to ascertain the success of the project from an instructor point of view:

- How is it going so far?
- How are the exercises working?
- What else do you need to teach this course well?

Responses were captured by question and by instructor and then categorized by common topics. For each topic, we noted how many instructors either expressed a similar point or noted, "I agree with that." Topics were then ordered from most to least frequently expressed. Topics are summarized in Table 4.

The fall semester beta instructors gave a positive review of the beta version, citing "easy to follow" exercises and materials, positive student performance on assignments and quizzes, and learning-focused teaching methodology as key indicators of success. Identified problems included technological issues, instructor-facing issues (such as unclear assignment specifications and inconsistent implementation of course assignments), and student-focused issues (such as lacking conceptual understanding and lacking motivation).

Instructors brainstormed how to resolve technology problems such as getting software set up on computers early in the course, getting students connected

Table 4 Summary of faculty feedback from fall beta test

1. How is it going?	Total
What is working well?	
Easy to follow exercises and materials	3
Student performance	3
Teaching methodology	2
What problems exist?	
Technological problems	3
Assignment specifications	2
Implementation issues	2
Understanding concepts	2
Motivation issues	2
2. How are the exercises working?	Total
Successful outcomes	
Building required skills	2
Preparing students for follow-on activities	2
Add more of the exercises	
Some topics require more practice	1
3. What else do you need?	Total
Additional materials	
Grading rubrics for some assignments	2
Structural additions	
Recommendations on taking attendance	3
Interior links to online text	2
Time estimates for instructors	2
Maintaining exam integrity	1
Process	
How to suggest course edits on a regular basis	1

sooner to the online text, and setting access correctly within software. They quickly assigned duties for fixing instructor-facing issues related to assignment specifications (e.g., requiring larger screenshots and a better naming convention for files) and implementation issues (e.g., setting appropriate permissions for graders and providing feedback more quickly to students).

All agreed that the student-focused issues were not specific to this particular course. They acknowledged that in all courses, students struggle with topics that are difficult to understand (e.g., in the FDB course: referential integrity and ER diagrams). Similarly, the motivational issues that they witnessed (e.g., some students not seeking help when needed or not completing assignments by the designated due date) were common.

When asked specifically about the in-class exercises, instructors offered only positive comments. They shared that in-class exercises help students to build required skills and that exercises prepare students well for the labs that follow next in the course sequence. Their only concern about in-class exercises was that they wanted more of them for some topics that were more difficult for students.

Instructors stated three things that would help them to teach the course better: the addition of grading rubrics for some assignments, a few structural additions to the course (e.g., recommendations for taking attendance, interior links to the online text, time estimates for in-class activities, and attention to exam security), and a process for noting changes to improve materials on an ongoing basis. By the end of the meeting, suggestions were assigned to an appropriate person who agreed to make the requested changes.

In addition to the early successes of the new FDB course learning outcomes and the early successes of the new FDB course instructional materials, the fall beta feedback session highlighted early successes in achieving shared understanding among instructors. At the meeting, faculty shared common instructional experiences, analyzed what they and their students needed, and volunteered to take responsibility for upgrades to course materials. Initial signs of consensus, community, and commitment emerged from sharing their experiences.

Follow-on Faculty Feedback Session One Year After Beta Test

After the beta version had been updated, we invited the nine instructors who taught the revised beta course in the following spring and fall semesters to participate in a faculty feedback session to debrief the experience and provide suggestions for the next course iteration. The first author of this chapter did not attend the meeting, so that instructors might feel more freedom to express their opinions.

Of the nine, two were assigned to the UP campus and had been part of the beta course since its inception. Both attended the meeting in person. Four were assigned to the online campus, and two of them were a part of the beta course since its inception. Two attended the meeting in person (only one of whom had been involved from the beginning), and two provided feedback in writing. Three instructors taught the course at small, regional residential campuses; and one of them provided feedback in writing. We received no response from the other two instructors from the small, regional residential campuses. A summary of instructor campus affiliations, semesters of teaching the beta course, and meeting participation appears in Table 5.

We asked four questions to gauge project success from an instructor point of view:

- In what ways do you think students benefit from the specification of this course?
- What is your assessment of the teaching and learning materials in this course?
- Describe the community aspect of being a "specified FDB course" instructor.
- How could this course be improved?

Responses were grouped by similarity of topics. In some instances, instructors named the topic. In other cases, the second author of this chapter named the topic. For each topic, we noted how many instructors either expressed a similar

Table 5 Campus affiliations and feedback meeting participation of instructors who taught the revised beta course in spring 2023 and/or fall 2023

			Spring 2023		Fall 2023	
Alias	Home campus	Mode of participation	RI	Online	RI	Online
A	Online	Written				
B	Large residential	In-person	X			
F	Small residential	Written			X	
G	Online	Written		X		X
J	Online	In-person				X
P	Online	In-person		X		X
R	Large residential	In-person	X		X	
O	Small residential	No response	X		X	
S	Small residential	No response	X			

point or noted, "I agree with that." Topics were then ordered from most to least frequently expressed within two major categories: strengths and weaknesses. Data are summarized in Table 6.

The two most notable strengths were (1) the structure of the course and its assignments and (2) the coordination among instructors. Many instructors agreed and no instructors disagreed with a major change in the course redesign: reordering the database design and SQL portions of the course, such that SQL was taught before design. Instructors who raised this point noted that this reversed sequence gave students a language to work with before learning to design a database. Two instructors who taught the course prior to the beta version explained that students in the beta version were more confident and prepared to address database design.

Instructors were equally pleased with the quality of the assignment materials, the clarity of the learning objectives, and the pacing of assignments. One online instructor said that the quality of assignments and how they are specified is particularly important for online students, who can work at their own pace. With the clear assignments and expectations, these students were able to make progress more easily than were students in the past. Instructors also praised the logical progression from in-class practice exercises to lab assignments to exams. They expressed that students also noted the usefulness of the instructor demo first, class practice second, and individual practice third progression of course activities that build from simple to moderate to advanced knowledge for each lesson.

The second most expressed strength, coordination among instructors, recurred throughout the discussion. Instructors commented that working with shared materials in a community of instructors saved them time given the many hands that made the workload lighter. They remarked that the time saved allowed them to spend more time with students who struggled to understand course concepts.

Also high on the list of strengths was consistency. Instructors shared positive outcomes of the consistency in course materials: better tutoring support because tutors know what all students are doing week to week, upper division course success because students were all similarly prepared, and predicted job market

Table 6 Summary of fall 2023 responses from faculty feedback session—in person and in writing responses combined

Strengths	Total
Structure of course and assignments	
Sequence—begin with SQL and follow with design	3
Quality, clarity, and pacing of materials	3
Logical progression of assignments	2
Coordination	
Coordination and community are key	8
Consistency	
Consistency of materials	5
General satisfaction	
Course and materials are good	3
Students provided positive feedback	2
Pedagogy	
Appropriate to build knowledge and complexity	3
Students understand the pedagogy	1
Areas for improvement/action items	Total
Assessment related	
Integrity of exams and quizzes	4
Easier questions for reading quizzes	3
Reduce grading workload	2
Assignment related	
Alter assignment format in the learning management system	3
Add a course change log	3
Technology related	
Support for online students' technical issues	2
Other	
Miscellaneous comments	3

success because of students' success in learning from consistent and well-developed materials.

Instructors expressed general satisfaction with the course materials–from both their vantage point and the vantage point of students. They also emphasized the appropriateness of the pedagogy underlying the course structure and said that students understood the rationale for the pedagogy once they progressed a few weeks into the course.

Instructors discussed three general areas for improvement and several specific suggestions for particular assignment elements. These points highlight the specific areas needing attention and enhancement.

The first area for improvement centered on assessment. Instructors expressed concern about the integrity of exams and quizzes. Especially with the proliferation of generative artificial intelligence tools, instructors discussed methods for securing integrity–and divided the work to do so. Regarding course reading quizzes, instructors recommended slightly easier questions, given that the purpose of these assessments is simply to establish familiarity with the concepts in the reading prior to class. And finally, instructors brainstormed ways to reduce the volume of grading. The top suggestion involved reducing the number of screenshots for instructors to review and verify.

The second area for improvement focused on the documentation of course assignments and changes to them. Instructors requested a different format for presenting assignments–a format that would be more easily editable by all instructors. Like the fall 2022 focus group, the instructors in the fall 2023 focus group also requested a change log for noting recommended updates as the course progressed—the only item that was not addressed from the earlier feedback meeting.

The third area for improvement involved technology. Instructors of online students requested information on supporting their students' technical issues (e.g., setting up SQL software, working on Mac). After the meeting, these instructors received the information that they sought.

The miscellaneous comments included "considering a refresher script to address 'drift' in datasets" and changing a few data points in the dataset. The first author responded to the individual instructors who offered these suggestions and determined which would significantly improve the course.

As a result of this and previous interactions, faculty members shared their successes, their concerns, and their particular implementation of the FDB course. Instructors whose students did not fare as well in some areas learned from instructors whose students succeeded in those areas. Consensus, community, and commitment increased in the sharing of experiences.

Survey of Campus Instructors Who Requested the Specified Course Materials

In fall 2023, we received a list of seven instructors who had requested to review the FDB course materials between the publication of the first beta version and the second. Since many of these instructors are located at regional campuses, we wanted to know if the course served the needs of those instructors and their students. We asked the seven instructors if they adopted all, some, or none of the specified course. We asked about their experience with the materials or about their reasons for not adopting the materials.

Four instructors responded to the survey. All four either used the materials, planned to use the materials in an upcoming semester, or referred the materials to a colleague. Two instructors elected to adopt the course "as is" and two elected to adopt parts of the course. Two instructors rated the materials as "exceptional." Two selected "neither good nor bad" or "I have not used them yet," as they had either not taught the course yet or were referring the materials to a colleague. While only one instructor shared concerns in the survey (i.e., considering how to run the course with students who don't own a laptop), a few campus instructors wrote separately

Table 7 Summary of fall 2023 responses from campus faculty who previewed and/or used the FDB materials

Course material adoption (all or parts)	Number of semesters using course materials	Overall assessment of course elements	Feedback on the course
All	0—but I plan to use them in the near future	Exceptional	I have not taught it with the SQL design approach before
All	1	Exceptional	I directed our adjunct who taught the course to use the materials because it was clearly the best
Parts	0—but I plan to use them in the near future	NA—I have not used them yet	I need to work out issues for students whose laptops don't support MySQL
Parts	1	Neither good nor bad	I recommended the materials to an adjunct instructor

to express concerns about teaching SQL before teaching design. See Table 7 for a summary of survey data.

Discussion of Findings

The findings suggest that the redesigned course was both relevant and well-received by students and participating faculty. In addition, data indicate that we reached our goal of achieving a shared understanding among faculty.

Success of Course Materials and Structure from a Design Perspective, a Student Perspective, and an Instructor Perspective

From a design perspective, the course met the three design elements identified in the literature review. First, the course employed an active learning strategy. Lectures were minimal to nonexistent, depending on the topic. Students came prepared (via pre-class assignments) to engage with the content, the instructor, and their peers during class. They actively worked to complete in-class exercises with the assistance of the instructional team. All instructors were pleased with student engagement, and students appreciated hands-on activities.

Second, goal-directed practice was the cornerstone of each lesson. The pre-, during-, and post-class activities were designed to advance students' knowledge of a given topic from introductory, to intermediate, to advanced. Each set of exercises was anchored in learning objectives that contributed to the overarching learning goals of the course. Both instructors and students commented on the usefulness of this structure and the intentional practice activities. Instructors reported that students expressed appreciation for the logical progression of assignments, which helped students build on knowledge from one topic to the next.

Third, students received ample feedback on their progress. The course design allowed for ongoing feedback opportunities for in-class exercises as well as two feedback opportunities on the practice exercise labs. While instructors recognized the value of the feedback, they discussed the need to reduce the amount of grading in the course. Together, they determined how to lessen the time spent on grading while keeping the valuable feedback elements. Simply removing some of the requirements for screenshots would accomplish the goal.

Students' evaluation of course materials was positive. They reported that the instructional materials were effective, the expectations for learning were clear, and the course content was relevant. They rated the in-class exercises, unit quizzes, and labs as highly beneficial and relevant to their learning. In addition, students performed well on the common exam that assessed the major learning objectives of the course. From the student perspective, the redesign was successful.

Faculty who participated in focus groups and those who responded to our survey also reported the effectiveness of the course structure and materials. Most faculty indicated that the structure of the course (i.e., teaching SQL prior to database design), the logical progression of the course assignments (i.e., from pre-class exercises to in-class exercises to post-class lab exercises), and the quality of the assignments were superior to previous versions of the course and promoted achievement of the stated learning objectives for the course.

Overall, the findings suggest that the redesigned FDB course is an effective way to teach students foundational database skills. These skills are relevant and immediately useful in the workplace, and students report securing internships and full-time employment based on their knowledge of SQL.

Shared Understanding

The redesigned FDB course was readily adopted by two campuses: UP and WC. Shared understanding was methodically established with these faculty. The WC faculty had spent time at UP as students; they met periodically throughout the beta test with UP faculty to discuss progress and concerns. Faculty helped one another across these locations. Cultural, geographic, and functional distance did not pose problems. The single UP faculty member who chose not to join the beta test (and thus did not adopt the redesigned course) is not teaching the course now. Thus, consistency was achieved across faculty who adopted the redesigned course.

Achieving shared understanding depended on three process elements: (1) clear vision for the course design by the first author, (2) inclusivity during the implementation of the beta version, and (3) consistent communication throughout the process. The first author's vision and leadership in design provided a defined set of materials from which instructors could work. This freed faculty members to make valuable contributions to collecting data, revising materials, adding assignments where needed, and generally advancing the course from one semester to the next–all of which promoted a shared understanding.

Inviting all instructors across the two target campuses (UP and WC) to opt in with the beta version allowed faculty choice to participate. All who accepted the invitation were committed to the process of testing and improving the course. Their buy-in led to sustained engagement in the community of FDB course instructors and further deepened their shared understandings.

Ongoing communication afforded cross-section coordination. Throughout the semester, instructors responded to email conversations, participated in regular check-in meetings, and shared student data. Their frequent connections contributed further to shared understanding.

Several regional campuses have adopted the redesigned FDB course, while others have asked to review the course materials. Various regional campus faculty have voiced concerns over aspects of the FDB course. These include use of the MySQL Server platform (as opposed to MS Access), challenges installing MySQL on Mac computers, questioning the sequence of database design coming after SQL components, and concerns with the amount of grading.

There is work to be done to increase course adoption across regional campuses. An approach we are trying is to pair an experienced FDB faculty with a new-to-the-FDB course faculty member. Next semester, an experienced FDB faculty from UP, who has previously taught the course in the traditional design-first style, is mentoring a regional campus faculty who voiced concerns over the course sequence. This happened organically. Hopefully, if we can get success stories from regional faculty, course adoption will increase. Currently, it is taught at three regional campuses.

Conclusions

The course development process was a collaborative effort led by the first author, who had a clear vision for the teaching approach and course structure. Weekly meetings were held with the two student assistants, both of whom took the FDB course before the redesign. This fostered a student-inclusive design process. Labs and the semester project were strategically built on weekly class exercises, creating a cohesive and progressive learning experience. The first author designed exercises, while student assistants contributed to labs and the end-of-semester project with guidance from the first author.

The beta test marked successful outcomes, with faculty members embracing the emphasis on SQL over design without the need for a hard sell. The well-thought-out course content and design, along with attention to detail, minimized problems for both faculty and students. The positive experience during the beta test contributed to the course's success, and faculty found it worthwhile for teaching and student learning.

While challenges exist in increasing course adoption at some regional campuses, with a few instructors expressing reluctance for assorted reasons, the first author is actively addressing concerns. In addition, a learning designer played a critical role in packaging the course for wider adoption, contributing to its accessibility and effectiveness. Early, respected, and vocal adopters continue to play a pivotal role in endorsing and promoting the course at key regional campuses and at University Park. Updates and presentations at all-state meetings are generating further engagement and dissemination of successful practices. All of these efforts contribute to a shared understanding of the need for consistent and relevant learning outcomes in the foundational FDB course.

References

Allen, D. E., Donham, R. S., & Bernhardt, S. A. (2011). Problem-based learning. *New Directions for Teaching and Learning: Evidence-Based Teaching, 128*, 21–29. Winter. https://doi.org/10.1002/tl.465

Ambrose, S. A., Bridges, M. W., DiPietro, M., Lovett, M. C., & Norman, M. K. (2010). What kinds of practice and feedback enhance learning? *How Learning Works, 7*, 121–152. Research-based Principles for Smart Teaching. Jossey-Bass.

Anders Ericsson, K. (2008). Deliberate practice and acquisition of expert performance: A general overview. *Academic Emergency Medicine, 15*, 988–994. https://doi.org/10.1111/j.1553-2712.2008.00227.x

Ausubel, D. P. (1963). *The psychology of meaningful verbal learning*. Grune & Stratton.

Bittner, E. A. C., & Leimeister, J. M. (2014). Creating shared understanding in heterogeneous work groups: Why it matters and how to achieve it. *Journal of Management Information Systems, 31*(1), 111–144.

Black, P. & William, D. (1998). Assessment and classroom learning. *Assessment in Education, 5*(1), 7–74. https://doi.org/10.1080/0969595980050102

Bonwell, C. C., & Eison, J. A. (1991). Active learning: Creating excitement in the classroom. In *ASHE-ERIC Higher Education Report No. 1*. Washington, D.C.: The George Washington University, School of Education and Human Development. https://files.eric.ed.gov/fulltext/ED336049.pdf

Bunderson, J. S., & Sutcliffe, K. M. (2002). Comparing alternative conceptualizations of functional diversity in management teams: Process and performance effects. *Academy of Management Journal, 45*(5), 875–893.

Chickering, A. W., & Gamson, Z. F. (1987). Seven principles for good practice in undergraduate education. *AAHE Bulletin, 3*, 7. https://files.eric.ed.gov/fulltext/ED282491.pdf

Cross, K. P. (1987). Teaching for learning. *AAHE Bulletin, 39*(8), 2–6. https://eric.ed.gov/?id=ED283446

Crouch, C. H., & Mazur, E. (2001). Peer instruction: Ten years of experience and results. *American Journal of Physics, 69*(9), 970–977.

Day, J. A., & Foley, J. D. (2006). Evaluating a web lecture intervention in a human–computer interaction course. *IEEE Transactions on Education, 49*(4), 420–431. https://doi.org/10.1109/TE.2006.879792

Espinosa, J. A., Cummings, J. N., Wilson, J. M., & Pearce, B. M. (2003). Team boundary issues across multiple global firms. *Journal of Management Information Systems, 19*(4), 157–190.

Fagen, A. P., Crouch, C. H., & Mazur, E. (2002). Peer instruction: Results from a range of classrooms. *The Physics Teacher, 40*, 206–209.

Foertsch, J., Moses, G., Strikwerda, J., & Litzkow, M. (2002). Reversing the lecture/homework paradigm using eTEACH web-based streaming video software. *Journal of Engineering Education, 91*(3), 263–365. https://doi.org/10.1002/j.2168-9830.2002.tb00703.x

Freeman, S., Eddy, S. L., McDonough, M., Smith, M. K., Okoroafor, N., Jordt, H., & Wenderoth, M. P. (2014). Active learning increases student performance in science, engineering, and mathematics. *Proceedings of the National Academy of Sciences of the United States of America (PNAS), 111*(23), 8410–8415. https://doi.org/10.1073/pnas.1319030111

Glick, W, H., Miller, G. G., & Huber, G, P. (1993). The impact of upper-echelon diversity on organizational performance. In G. P. Huber & W. H. Glick (Eds.), In *Organizational change and redesign: Ideas and insights for improving performance* (pp. 176–214). New York: Oxford.

Gregory, W. G., Beck, R., & Keil, M. (2013). Control balancing in information systems development offshoring projects. *MIS Quarterly, 37*(4), 1211–1232.

Hake, R. R. (1998). Interactive-engagement vs. traditional methods: A six-thousand-student survey of Mechanics test data for introductory physics courses. *American Journal of Physics, 66*, 64–74.

Halpern, D. F., & Hakel, M. D. (2003). Applying the science of learning to the university and beyond. *Change, 37*, 37–41.

Hattie, J. (2009). *Visible learning: A synthesis of over 800 meta-analyses relating to achievement*. Routledge.

Hattie, J. (1992). Measuring the effects of schooling. *Australian Journal of Education, 36*(1), 5–13.

Hinds, P., & Mortensen, M. (2005). Understanding conflict in geographically distributed teams: An empirical investigation. *Organization Science, 16*(3), 290–307.

Hung, W., Jonassen, D. H., & Liu, R. (2008). Problem-based learning. In J. M. Spector, M. D. Merrill, J. Van Merriënboer, & M. P. Driscoll (Eds.), *Handbook of research on educational communications and technology* (3rd ed., 485–506).

Jenkin, T. A., Chan, Y. E., & Sabherwal, R. (2019). Mutual understanding in information systems development: Changes within and across projects. *MIS Quarterly, 43*(2), 649–671.

Jonsson, A., & Svingby, G. (2007). The use of scoring rubrics: Reliability, validity and educational consequences. *Educational Research Review, 2*(2), 130–144. https://www.sciencedirect.com/science/article/abs/pii/S1747938X07000188

Kaddoura, M. (2013). Think Pair Share: A teaching learning strategy to enhance students' critical thinking. *Educational Research Quarterly, 36*(4), 3–24.

Kothiyal, A., Murthy, S., & Iyer, S. (2014). Think-pair-share in a large CS1 class: Does learning really happen? ResearchGate. https://doi.org/10.1145/2591708.2591739

Leinonen, J., Denny, P., & Whalley, J. (2022). A comparison of immediate and scheduled feedback in introductory programming projects. In *SIGCSE 2022: Proceedings of the 53rd ACM technical symposium on computer science education* (Vol. 1, pp. 885–891). https://doi.org/10.1145/3478431.3499372

Malik, K. M., & Zhu, M. (2023). Do project-based learning, hands-on activities, and flipped teaching enhance student's learning of introductory theoretical computing classes? *Education and Information Technologies, 28*, 3581–3604. https://doi.org/10.1007/s10639-022-11350-8

Martins, E. C. (2003). Building organizational culture that stimulates creativity and innovation. *European Journal of Innovation Management, 6*(1), 64–74.

Michaelsen, L. K., & Sweet, M. (2008). The essential elements of team-based learning. In *New directions for teaching and learning: Team-based learning: small group learning's next big step* (Vol. 116, pp. 7–27). Winter. https://doi.org/10.1002/tl.330

Mohammed, S., & Dumville, B. C. (2001). Team mental models in a team knowledge framework: Expanding theory and measurement across disciplinary boundaries. *Journal of Organizational Behavior, 22*, pp. 89–106.

Novak, J. D. (2002). Meaningful learning: The essential factor for conceptual change in limited or inappropriate propositional hierarchies leading to empowerment of learners. *Science Education, 86*(4), 548–571. https://doi.org/10.1002/sce.10032

Oakley, B. (2014). *A mind for numbers: How to excel at math and science*. The Penguin Group.

Ocker, R. J., Huang, H., Benbunan-Fich, R., & Hiltz, S. R. (2011). Leadership dynamics in partially distributed teams: An exploratory study of the effects of configuration and distance. *Group Decision and Negotiation, 20*, 273–292.
Ocker, R. J., Rosson, M. R., Kracaw, D., & Hiltz, S. R. (2009). Training students to work effectively in Partially Distributed Teams. *ACM Transactions on Computing Education, 9*(1), 1–24.
Roschelle, J. (1995). Learning in interactive environments: Prior knowledge and new experience. In J. H. Falk & L. D. Dierking (Eds.), *Public institutions for personal learning: Establishing a research agenda* (pp. 37–51). Washington, DC: American Association of Museums.
Slavin, R. E. (1996). Research on cooperative learning and achievement: What we know, what we need to know. *Contemporary Educational Psychology, 21*, 43–69.
Sutcliffe, K. M. (1994). What executives notice: Accurate perceptions in top management teams. *Academy of Management Journal, 37*, 1360–1378.
Weiman, C. (2014). Large-scale comparison of science teaching methods sends clear message. *Proceedings of the National Academy of Sciences of the United States of America (PNAS), 111*(23), 8319–8320. https://doi.org/10.1073/pnas.1407304111
Wilson, S. G. (2013). The flipped class: A method to address the challenges of an undergraduate statistics course. *Teaching of Psychology, 40*(3), 193–199. https://doi.org/10.1177/0098628313487461

The manufacturer's authorised representative in the EU is Springer Nature Customer Service Centre GmbH, Europaplatz 3, 69115 Heidelberg, Germany. If you have any concerns regarding our products, please contact ProductSafety@springernature.com

Printed and bound by CPI Group (UK) Ltd, Croydon, CR0 4YY
05/10/2026
02227724-0002